The Gen AI Blueprint

First Edition

The content of GenAI Blueprint provides readers with a comprehensive framework for implementing and scaling Generative AI solutions, offering practical strategies, architectural guidelines, and real-world examples. It equips professionals with the tools and best practices needed to navigate the complexities of building secure, efficient, and scalable AI systems in today's evolving landscape.

Shyam Narayan

The Gen AI Blueprint

GenAI was instrumental in shaping the insights and content for 'GenAI Blueprint,' providing valuable assistance in researching, structuring, and refining complex concepts into a comprehensive guide for AI enthusiasts and professionals alike.

Thank you, Gen AI technologies, for helping me author this book. This book would not be possible without evolution of GenAI technologies

- Shyam Narayan

Preface

In the ever-evolving landscape of artificial intelligence, **Generative AI** stands as a transformative force, reshaping industries, redefining innovation, and challenging traditional norms. As this groundbreaking technology becomes more accessible, its applications continue to expand—from generating human-like text and realistic images to enabling highly personalized user experiences across domains. Yet, with this potential comes complexity, and navigating the intricacies of Generative AI requires a clear roadmap.

The GenAI Blueprint is designed to be that roadmap. Authored by **Shyam Narayan**, this book combines decades of professional expertise with the power of Generative AI technologies to deliver a comprehensive guide that demystifies the end-to-end journey of Generative AI—from foundational concepts to advanced implementations. This book is not just about the technology; it's about how to harness its power responsibly, effectively, and ethically.

What to Expect in This Book

The **GenAI Blueprint** is a structured and practical guide that covers the entire lifecycle of Generative AI. Here's what you'll find in this book:

1. A Historical Perspective on AI

The book begins with **Chapter 1: Evolution of AI**, tracing the journey from early rule-based systems to today's Generative AI breakthroughs. This historical context provides readers with a solid foundation for understanding how AI has evolved and where it is headed.

2. A Comprehensive End-to-End Process Flow

In **Chapter 2**, the book delves into the **Generative AI End-to-End Process Flow**, offering a step-by-step exploration of the key stages involved in building Generative AI solutions. Readers will gain insights into the practical workflow, from data collection to deployment and monitoring.

3. Deep Technical Exploration of LLM Architectures

Chapters 3 and 4 focus on the heart of Generative AI—Large Language Models (LLMs).

- In **Chapter 3**, you'll explore various **LLM architectures**, their design principles, and the trade-offs between different approaches.
- **Chapter 4** provides guidance on **choosing the right LLM** for specific use cases, helping readers make informed decisions based on performance, scalability, and cost.

4. Design and Architectural Principles

Chapters 5 and 6 cover the **Design and Architecture Principles for Generative AI Solutions** and provide **reference architectures** that serve as practical blueprints for building scalable, modular, and secure AI systems. These chapters blend theoretical principles with actionable insights and visual aids.

5. Advanced Topics in Security and Data Strategy

In **Chapters 7 and 9**, the book addresses the critical aspects of security and data readiness:

- **Chapter 7: Security Considerations** dives into safeguarding Generative AI systems against vulnerabilities, including adversarial attacks and data breaches.
- **Chapter 9: Evolving the Data Strategy for Generative AI Maturity** emphasizes the importance of scalable and real-time data pipelines to support advanced AI systems.

6. Practical Techniques for Prompt Engineering

In **Chapter 8**, you'll learn about **Prompt Engineering**, a vital skill for optimizing AI outputs. This chapter includes tips, examples, and best practices for crafting effective prompts to elicit desired responses from Generative AI models.

7. Comprehensive Testing and Operationalization

Chapters 10 and 11 focus on the **testing strategies** and the relevance of **MLOps** for Generative AI implementations. These chapters provide methodologies and tools to ensure model robustness, reliability, and seamless integration into production systems.

8. Governance and Ethical Considerations

In **Chapters 12, 14, and 15**, the book explores the ethical dimensions of Generative AI:

- **Chapter 12: Role of Governance in Generative AI** highlights frameworks for responsible AI usage.
- **Chapter 14: Responsible AI** emphasizes trust and accountability in deploying AI solutions.
- **Chapter 15** addresses the **misuse of Generative AI,** discussing risks and strategies to safeguard against potential threats.

9. Real-World Use Cases and Industry Applications

Chapters 16 and 18 showcase the transformative applications of Generative AI:

- **Chapter 16: Generative AI Risk Categorizing** examines risks across industries, offering mitigation strategies.
- **Chapter 18** provides a **comprehensive list of use cases**, illustrating how Generative AI is revolutionizing sectors like healthcare, finance, retail, and media.

10. Leadership in the Generative AI Era

Chapter 17: The Role of the Chief Generative AI Officer outlines the responsibilities of this emerging leadership role, offering insights into how organizations can align AI initiatives with business goals.

Who Should Read This Book

This book is for:

- **AI Practitioners and Developers:** Looking for technical guidance on building and deploying Generative AI solutions.
- **Enterprise Leaders and Strategists:** Interested in understanding the strategic and operational aspects of leveraging Generative AI for business growth.
- **Researchers and Educators:** Seeking a structured overview of Generative AI technologies and methodologies.
- **Students and Enthusiasts:** Exploring the potential of Generative AI and its practical applications.
- **Security and Governance Professionals:** Focused on ensuring safe and ethical use of Generative AI systems.

Whether you are a technical expert, a decision-maker, or simply curious about the possibilities of Generative AI, this book provides insights tailored to your needs.

About the Author

Shyam Narayan is a thought leader and seasoned technologist with extensive experience in solutioning and implementing complex AI transformations across industries. As a leader, Shyam is passionate about leveraging AI and cloud technologies to create value for clients. This book reflects his deep expertise, practical insights, and commitment to responsible AI practices. Written with the assistance of Generative AI technologies, the book itself exemplifies how AI can augment human creativity and productivity.

How to Use This Book

This book is structured to cater to readers with varied objectives:

- If you are **new to Generative A**I, start with foundational chapters such as "Evolution of AI" and "Overview of the Generative AI End-to-End Process Flow."
- For a **deep dive into technical implementations**, focus on chapters covering architecture design, LLM creation, and reference architectures.
- If you are concerned about **security, governance, and ethical considerations**, explore chapters on security, governance, and Responsible AI.
- For **industry-specific insights**, refer to chapters on use cases and risks associated with Generative AI.

Each chapter can be read independently, making it a flexible resource for learning at your own pace or revisiting specific topics.

Acknowledgements

Writing this book has been a journey of collaboration, inspiration, and discovery. I extend my heartfelt gratitude to the incredible teams and the broader AI community for their unwavering support and insights.

I am particularly thankful to the developers and researchers who have pioneered Generative AI technologies, whose work laid the foundation for this book. Special thanks to the creators of Generative AI tools that assisted in streamlining this writing process, exemplifying the technology's transformative potential.

Lastly, my deepest appreciation goes to my family, colleagues, and mentors for their encouragement, patience, and belief in this endeavor.

Final Thoughts

As we stand at the intersection of technological promise and profound societal impact, it is our responsibility to ensure that Generative AI is used as a force for good. By fostering trust, accountability, and innovation, we can unlock its full potential to benefit all.

Welcome to **The GenAI Blueprint**. Let this be your guide to navigating the future of AI innovation.

Shyam Narayan

Author, The Gen AI Blueprint

November 2024

Foreward

Foreward By Phil Fersht,

CEO and Chief Analyst, HFS Research

Narayan articulates a pragmatic and intriguing evolution of AI to set the scene for the influx of Large Language Models. I especially enjoyed how he sets the context for the role of AI governance in organizations to manage the proliferation of AI technologies. The GenAI Blueprint is a must-read for enterprise technology and business executives to get to grips with the current state of AI and how to coach themselves and their enterprises to stay ahead of the curve.

Foreward By Niranjan Prabhu,

Chief Information Officer | Board Member | Top 50 Most Innovative CIOs 2023/2020

AI is no longer on the horizon—it is firmly embedded in our lives. Generative AI is already redefining how we innovate, create, and tackle complex challenges. This transformation goes beyond mere technological advancements; it is reshaping how we think about productivity, decision-making, and value creation, unlocking possibilities that we are only beginning to explore.

However, with this extraordinary potential comes significant responsibility. How can organisations adopt this technology effectively, securely, and ethically? How can they balance the transformative capabilities of Generative AI with the need for accountability and trust? These are the critical questions addressed in The Gen AI Blueprint, a comprehensive and practical guide for navigating the complexities of Generative AI.

This book is both timely and essential, offering a structured roadmap for organisations eager to unlock the potential of Generative AI. It provides invaluable guidance for leaders, practitioners, and innovators, covering everything from the evolution of artificial intelligence to the design of Large Language Model (LLM) architectures, security frameworks, and governance strategies.

What sets this work apart is its holistic approach. It does not stop at technical exploration but delves deeply into overlooked areas, such as prompt engineering, MLOps, and responsible AI practices. The dedicated chapters on ethical considerations and risk mitigation serve as a crucial foundation for building AI systems that are both effective and trustworthy.

Practicality is at the heart of this blueprint. The inclusion of industry-specific use cases and the emerging role of the Chief Generative AI Officer ensures its relevance across a broad spectrum of applications. Whether you are developing scalable systems, fine-tuning AI interactions, or leading strategic transformation, this guide offers actionable insights and proven best practices to help you move beyond the hype and achieve meaningful outcomes.

The Gen AI Blueprint is more than just a technical manual; it is a call to action for leaders to navigate this transformative period with wisdom and responsibility. It equips readers with the tools, frameworks, and foresight needed to harness the immense power of Generative AI while addressing its inherent challenges with clarity and purpose.

The path forward is as exciting as it is complex. With The Gen AI Blueprint, you are well-prepared to lead your organisation into the future with confidence, integrity, and strategic clarity.

Foreward By Guneet Singh

Transformational Technology Leader; Alum-Google, Microsoft, Samsung

I often wonder—what would humanity's reaction to AI have been if we had called it Assistive Intelligence instead of Artificial Intelligence? Would we have embraced it more readily as a collaborator rather than questioning its intentions? As someone who has led technology (and recently AI!) adoption across global organisations like Google, Microsoft, and Samsung, I've witnessed firsthand the varied responses to this transformative technology. I've witnessed a spectrum of reactions. From awe-inspiring acceptance among consumers, to cautious curiosity in office environments, to strategic (and, often challenged) deployment within enterprises - one thing is clear: AI's greatest promise lies in its ability to assist us.

This brings me to Shyam Narayan's The Gen AI Blueprint, a book that truly puts the "assisting" in AI. This is not just another conceptual discourse on Generative AI. Instead, it is a practical guide packed with actionable insights and real-world examples. Drawing on his extensive two-decade career in technology deployment, Shyam demystifies the myriad acronyms, architectures, and tools in the AI landscape, transforming them into an accessible roadmap for building impactful AI solutions.

What sets this book apart is its seamless journey across the many dimensions of Generative AI. From the technical core of Chapter 6, where Shyam expertly dissects Generative AI Architectures, to the user-level focus of Chapter 8 on Prompt Engineering, and even the organisational foresight in Chapter 16 on the emergence of the Chief Generative AI Officer, this book captures both concrete architectural blueprints and deployment strategies helping you chart a successful journey with AI.

Shyam's ability to simplify the complex shines throughout the book. Each chapter is brought to life with relatable case studies, bridging theory and practical application. Whether he's guiding readers on deploying scalable AI solutions or preparing organisations for the human-centric challenges of AI adoption, Shyam's insights reflect a deep understanding of what it takes to bring transformative technology to the real world.

I've known Shyam since our days together in business school. I fondly remember his patient yet effervescent manner of explaining concepts over a cup of tea on campus. No matter how complex the topic, Shyam had a way of cutting through the noise, making the hardest ideas easy to understand. He was remarkable, and this book takes me right back to those moments. I can almost imagine him now, thoughtfully and enthusiastically demystifying AI for his eager audience.

Congratulations, Shyam, on writing this exceptional book. It not only provides a comprehensive roadmap for navigating the world of Generative AI but also puts the much-needed "assist" into AI for its readers.

Foreward By David Fodor

Technology Transformation Leader

There are times in our lives when we are privileged to witness the emergence of new technology that has a profound impact on the way we live and work. We are in the middle of one of these globally impactful events right now. The reawakening of Artificial Intelligence (AI) through the emergence of Generative AI has ignited the interest of the general population and left us wondering of the seemingly boundless limits of its application.

The reality is that Artificial Intelligence is nothing new. It is a technology that has existed since the mid-20th century. What has changed, enabling the broad capabilities of Generative AI is the growing magnitude of data and information available to build and feed predictive models coupled with the immense power of modern, cloud-hosted computing.

This book comes at a critical time in the journey to monetise the latent potential of Generative AI. Over the past two years AI and Machine Learning practitioners have experimented with few precedents that provide guidelines and guardrails to ensure safe and efficient application of the technology. Shyam has created a comprehensive playbook that will help deepen the understanding of the technology itself whilst providing a practical set of steps to ensure optimal adoption.

Shyam has been working at the forefront of technology and transformation for over 25 years thus has had a front-row seat to witness the evolution of GenAI. I have enjoyed the pleasure of collaborating closely with Shyam over much of that time witnessing his thought leadership firsthand. Shyam is a deep thinker with the ability to create compelling visions of the future. It is therefore not surprising he has created a comprehensive handbook that will provide a mechanism to navigate successfully what remains an emerging technology.

We cannot foresee the evolution of GenAI, nor the use cases that will be prosecuted in the future leveraging the technology. However, what we can be certain of is that we require clear pathways to ensure the capture of full value from this ground-breaking intelligence. This book will help you to build those pathways, acting as your innovation toolkit. I hope that you gain from it as much value as I have!

Table of Contents

Chapter 1

Evolution of AI

The history of Artificial Intelligence (AI) is a rich tapestry of ideas, innovations, and breakthroughs that have redefined the boundaries of human ingenuity. From its humble beginnings as a theoretical concept to its current status as a transformative force across industries, AI has evolved through various stages, each marked by distinct technological paradigms and philosophical shifts. This chapter provides an in-depth look at the key milestones and breakthroughs that have shaped the evolution of AI over the decades.

1. Early Beginnings: The Birth of AI Concepts

AI's conceptual origins date back to the early 20th century, rooted in the ideas of mathematics, philosophy, and cognitive science. Thinkers began to ponder whether machines could replicate human reasoning, sparking curiosity and laying the groundwork for the field of AI. The period was characterized by a blend of philosophical speculation and early experimentation with the foundations of logic and computation.

- **Alan Turing and the Universal Machine (1936):** Alan Turing, a British mathematician, introduced the concept of a *Universal Machine*, which later became known as the *Turing Machine*. Turing's work established the theoretical foundations for what would eventually be called a computer. His 1936 paper, "On Computable Numbers," demonstrated that a machine could perform any conceivable mathematical computation if it were represented as an algorithm. This was the precursor to the idea that machines could "think" or "reason" if provided with the right set of instructions.
- **The Turing Test (1950):** In 1950, Turing published the seminal paper "Computing Machinery and Intelligence," where he posed the question, "Can machines think?" In this work, he proposed what is now known as the *Turing Test*. The test evaluates a machine's ability to exhibit intelligent behavior by engaging in a conversation that is indistinguishable from that of a human. If the human evaluator cannot reliably distinguish the machine from the human, the machine is said to have passed the test. This concept sparked both excitement and controversy, setting the stage for future debates about machine intelligence.

- **Dartmouth Conference (1956):** The Dartmouth Conference, organized by John McCarthy, Marvin Minsky, Nathaniel Rochester, and Claude Shannon, marked the official birth of AI as a field of research. At this conference, McCarthy coined the term *Artificial Intelligence*. The participants believed that "every aspect of learning or any other feature of intelligence can, in principle, be so precisely described that a machine can be made to simulate it." Although the expectations were high, progress was slow, and early efforts struggled to deliver on these lofty promises.
- **Early AI Programs:** Following the Dartmouth Conference, early AI programs were developed to explore specific aspects of intelligence. Notable examples include the *Logic Theorist* (1956) by Allen Newell and Herbert A. Simon, which could prove mathematical theorems, and *General Problem Solver (GPS)* (1957), which aimed to solve a wide range of problems using heuristic methods.

These early milestones set the conceptual and technical foundation for AI, framing it as a discipline that combined elements of computer science, mathematics, and cognitive psychology.

2. The Rise and Fall of Symbolic AI (1950s–1980s)

The first few decades of AI research were dominated by the notion that human intelligence could be replicated through symbolic manipulation. This period, often referred to as the era of *Good Old-Fashioned AI (GOFAI)*, focused on creating intelligent behavior by manipulating symbols and applying logical rules. Symbolic AI was rooted in the belief that cognitive processes could be represented as formal operations on abstract symbols.

- Rule-Based Systems and Early Achievements: Early AI systems were based on formal rules and logic. Researchers developed rule-based systems that encoded knowledge into explicit rules. For example, SHRDLU (1968), a natural language understanding program developed by Terry Winograd, could understand and respond to queries about blocks in a virtual world using pre-defined rules. SHRDLU was considered a major breakthrough at the time, demonstrating that machines could process language and reason about spatial relationships.
- **Expert Systems (1970s):** Symbolic AI reached its peak with the advent of *Expert Systems*. These systems used knowledge bases and inference engines to solve specific problems in narrow domains, such as medical diagnosis or chemical analysis. Notable examples include *MYCIN* (1972), an expert system for diagnosing bacterial infections, and *DENDRAL* (1965), used for chemical analysis. These systems could outperform human experts in very specialized tasks, leading to significant commercial interest.
- **Limitations of Symbolic AI:** Despite their early promise, symbolic AI systems struggled with real-world complexities. The *Frame Problem*, for instance, highlighted the difficulty of representing common-sense knowledge and handling dynamic, uncertain environments. As a result, these systems were brittle and unable to adapt to new situations outside their predefined knowledge base. This rigidity ultimately led to the decline of symbolic AI and the first *AI Winter* (1974–1980), a period marked by reduced funding and interest due to the lack of practical progress.

3. Emergence of Machine Learning (1980s–2000s)

The limitations of symbolic AI prompted a shift towards a new paradigm—*Machine Learning* (ML). Unlike symbolic AI, which relied on predefined rules, ML focused on *data-driven learning*. The core idea was that instead of programming a system with explicit instructions, one could enable it to learn patterns and behaviors from large datasets.

- **The Revival of Neural Networks:** Neural networks, inspired by the structure and function of the human brain, were initially proposed in the 1940s. However, it wasn't until the 1980s that a major breakthrough occurred with the development of the *backpropagation* algorithm. Geoffrey Hinton, David Rumelhart, and Ronald J. Williams introduced backpropagation as a method for training multilayer neural networks. This innovation allowed networks to learn complex mappings from inputs to outputs, making them suitable for tasks such as image recognition and language processing.
- **The Emergence of Statistical Learning (1990s):** During the 1990s, AI research embraced statistical methods. Techniques like Support Vector Machines (SVMs), Hidden Markov Models (HMMs), and Bayesian Networks gained prominence. These approaches, grounded in probability theory and statistics, allowed AI systems to handle uncertainty and learn from noisy, incomplete data. This period also saw the rise of *data mining*, where researchers focused on extracting patterns and insights from large datasets.
- **Applications and Early Successes:** Machine learning found early success in applications like speech recognition, handwriting analysis, and recommendation systems. For example, IBM's *Deep Blue* used advanced search algorithms and machine learning techniques to defeat world chess champion Garry Kasparov in 1997. This victory was seen as a symbolic triumph for AI, showcasing its ability to tackle complex problems that required strategic thinking.

The shift from rule-based systems to learning-based models laid the groundwork for the next revolution in AI: *Deep Learning*.

4. The Deep Learning Revolution (2010s)

The early 2010s marked the beginning of the *Deep Learning* revolution, which propelled AI into mainstream applications. Deep learning, a subset of ML, utilizes deep neural networks with many layers to automatically learn hierarchical representations of data. This approach proved to be highly effective for tasks like image recognition, natural language processing, and autonomous driving.

- **Breakthrough in Image Recognition (2012):** The breakthrough moment for deep learning came in 2012, when Geoffrey Hinton's team won the ImageNet competition—a large-scale visual recognition challenge—with a deep convolutional neural network (CNN) called *AlexNet*. AlexNet achieved a top-5 error rate of 15.3%, significantly outperforming traditional machine learning methods. This victory demonstrated the potential of deep learning for solving complex visual tasks.
- **Natural Language Processing (NLP) and Transformers (2017):** Another major leap occurred in the field of NLP with the introduction of *Transformers*. The Transformer architecture, introduced by Vaswani et al. in 2017, replaced traditional RNNs and LSTMs with a novel attention mechanism, enabling the training of much larger models. Transformers paved the way for large-scale language models like *BERT* (2018) and *GPT-2* (2019), which achieved state-of-the-art performance in language understanding and generation.
- **Wide Adoption and Real-World Impact:** Deep learning models began to outperform human-level accuracy in various tasks, such as facial recognition, speech synthesis, and even medical image analysis.

Companies like Google, Facebook, and Amazon integrated deep learning into their core products, from search engines to recommendation systems and voice assistants.

The deep learning revolution not only transformed AI research but also led to an explosion of AI-driven applications across industries.

5. AI in the Age of Generative Models (2020s–Present)

The 2020s have seen the rise of *Generative AI*, a field focused on creating new content—text, images, music, and more—using generative models. This era is characterized by powerful models that can generate human-like content, leading to new creative possibilities and applications.

- **Generative Adversarial Networks (GANs):** GANs, introduced by Ian Goodfellow in 2014, consist of two neural networks—a generator and a discriminator—that work in tandem. The generator creates new data samples, while the discriminator evaluates their authenticity. This adversarial training framework has been used to generate realistic images, art, and even synthetic data for training other models.
- **Transformer-Based Models:** The introduction of Transformer architectures led to the creation of large-scale language models like *GPT-3*, which can generate coherent and contextually accurate text based on prompts. These models are being used for everything from chatbots and content generation to code synthesis and virtual assistants.
- **Multimodal Models:** Recent advancements have seen the development of multimodal models like *DALL-E* and *CLIP*, which integrate multiple data types (text and images) to create visual content based on textual descriptions. These models represent a significant leap in AI's ability to understand and generate complex, creative outputs.

Generative AI has opened new frontiers in creativity and automation, enabling machines to not just analyze, but also *create*, blurring the lines between human and machine-generated content.

6. Current Trends and the Future of AI

The current era of AI is marked by rapid advancements and new challenges. Several key trends and research directions are shaping the future trajectory of the field:

- **Explainability and Ethics:** As AI systems become more complex, understanding how they make decisions has become crucial. Explainable AI (XAI) aims to provide insights into model behavior, making them more transparent and trustworthy. Ethical AI focuses on ensuring that these systems are fair, unbiased, and used responsibly.
- **AI at the Edge and Federated Learning:** Edge AI moves computation closer to data sources (e.g., IoT devices), enabling real-time, low-latency processing. Federated learning, on the other hand, trains models across decentralized devices without sharing raw data, preserving user privacy.
- **Artificial General Intelligence (AGI):** While current AI systems excel in narrow tasks, AGI aims to achieve human-level intelligence across a wide range of domains. Although AGI remains a distant goal, research in areas like transfer learning, reinforcement learning, and unsupervised learning continues to push the boundaries.

- **Quantum AI:** The convergence of AI and quantum computing promises to unlock new computational capabilities, solving problems that are currently intractable for classical computers.

Timeline View of AI Evolution

This timeline provides an overview of key developments in the evolution of AI. Let me explain some of the significant milestones and periods:

1. Early AI (1930s-1960s):
    - 1956: Dartmouth Conference, widely considered the birth of AI as a field
    - 1958: Perceptron, one of the first artificial neural networks
    - 1966: ELIZA, the first chatbot, demonstrating natural language processing
2. AI Winter & Revival (1970s-1990s):
    - 1974-1980: First AI Winter, a period of reduced funding and interest
    - 1980-1987: Rise of Expert Systems, rule-based AI for specific domains
    - 1987-1993: Second AI Winter, another period of setbacks
3. Modern AI (1990s-2010s):
    - 1997: Deep Blue defeats world chess champion Garry Kasparov
    - 2001-2010: Start of Big Data era, providing fuel for AI advancements
    - 2010-2017: ImageNet competition drives progress in computer vision
    - 2016: AlphaGo defeats world Go champion Lee Sedol
4. Deep Learning Revolution (2018-2020):
    - 2018: GPT-1 and BERT introduce transformer-based language models
    - 2019: GPT-2 shows impressive text generation capabilities
    - 2020: GPT-3 demonstrates few-shot learning and versatility
5. Recent Breakthroughs (2021-Present):
    - 2021: DALL-E combines language and image generation
    - 2022: ChatGPT brings conversational AI to the mainstream
    - 2023: GPT-4 shows significant improvements in reasoning and multimodal capabilities
    - 2023-2024: Rapid advancements in multimodal AI, combining text, image, and other data types

This timeline shows the cyclical nature of AI development, with periods of great progress followed by setbacks, and then renewed advancements. The field has accelerated dramatically in recent years, particularly with the rise of deep learning and large language models.

Conclusion

The evolution of AI is a story of relentless innovation and exploration, driven by humanity's quest to build machines that can think and learn. From symbolic logic to deep learning and generative models, each era has contributed to the foundation of what we now call AI. As AI continues to evolve, it will play an even greater role in shaping the future of technology and society, raising profound questions about the nature of intelligence and our relationship with machines

Chapter 2

Overview of the Generative AI End-to-End Process Flow

Generative AI has transformed from a niche research area into a mainstream technology that powers a variety of innovative applications, from text generation and image creation to complex decision-making systems. However, building and deploying a successful Generative AI solution is a complex, multi-stage process that requires meticulous planning, deep technical expertise, and a structured approach. This chapter provides a comprehensive overview of the **Generative AI End-to-End Process Flow**, detailing each step involved—from data preparation and model development to deployment, monitoring, and continuous optimization.

The goal of this chapter is to offer a structured roadmap that guides data scientists, engineers, and business leaders through the various stages of implementing a Generative AI solution. By breaking down the entire lifecycle into clear, actionable steps, this chapter aims to demystify the process and provide practical insights, best practices, and tools for each stage. Understanding this flow is essential for building scalable, reliable, and impactful Generative AI systems.

Why an End-to-End Process Flow is Essential

Given the complexity and unique challenges of Generative AI, it's critical to have a well-defined process that outlines the sequence of activities required to develop and deploy these models. Unlike traditional machine learning workflows, Generative AI involves additional steps such as managing large-scale pre-trained models, fine-tuning on domain-specific data, handling creative outputs, and integrating ethical considerations. An end-to-end process flow ensures:

- **Consistency and Repeatability**: Establishing a clear process flow reduces the risk of inconsistencies and errors, ensuring that the AI models deliver predictable and reliable results.
- **Efficiency and Scalability**: A structured workflow helps optimize resources, reduce development time, and scale the solution more efficiently.
- **Compliance and Governance**: Having a documented process flow is essential for compliance and traceability, especially in industries with stringent regulatory requirements.

This chapter will delve into each stage of the process, providing a detailed roadmap for building robust Generative AI models that can be deployed confidently in real-world applications. The lifecycle of a Generative AI project typically consists of the following key stages:

1. **Data Collection and Preparation**
2. **Model Selection and Architecture Design**
3. **Model Training**
4. **Evaluation and Validation**
5. **Fine-Tuning and Optimization**
6. **Deployment and Integration**
7. **Continuous Monitoring and Maintenance**
8. **Choosing the Right LLM for Specific Use Cases**

Each stage has its own unique set of challenges, requirements, and tools. This chapter will explore each stage in depth, providing a clear understanding of the best practices and strategies for building successful Generative AI models.

1. Step-by-Step Breakdown of the Generative AI Process

1.1. Data Collection and Preparation

Objective: Gather, clean, and prepare high-quality data for training the Generative AI model.

Detailed Process:

- **Data Collection**: Begin by identifying the type and sources of data required for your specific Generative AI use case. For example, a text generation model would require large-scale text datasets, such as news articles, research papers, or social media posts, whereas an image generation model might require labeled images with metadata.
  - **Text Data Sources**: Wikipedia, Common Crawl, news websites, and proprietary company documents.
  - **Image Data Sources**: Open Image Dataset, COCO Dataset, and proprietary image repositories.
- **Data Cleaning**: Once data is collected, the next step is cleaning it to remove inconsistencies, irrelevant information, and errors. Techniques include removing duplicates, handling missing values, filtering out noisy or low-quality data, and normalizing text for NLP tasks.
- **Data Labeling**: In supervised learning, data labeling is crucial. For instance, in a sentiment analysis use case, each text snippet needs to be labeled as positive, negative, or neutral. Similarly, image data might require bounding boxes, segmentation masks, or class labels.

- **Data Transformation**: For Generative AI, transforming data into a suitable format (e.g., converting images to tensors or tokenizing text data) is necessary. Text data, for example, must be tokenized and structured appropriately.

Example: For a text summarization model, you might collect a dataset of 500,000 news articles along with their corresponding human-written summaries. This dataset will serve as the training and validation set for the Generative AI model.

Best Practices:

- **Ensure Data Diversity**: Include a variety of sources to prevent overfitting and ensure that the model generalizes well.
- **Automate Data Cleaning**: Implement automated pipelines for repetitive cleaning tasks using tools like **AWS Glue**, **Google Cloud Dataprep**, or **Azure Data Factory**.
- **Document Data Lineage**: Maintain a record of data sources, transformations, and labeling rules for reproducibility.

1.2. Model Selection and Architecture Design

Objective: Choose an appropriate model architecture based on the complexity and requirements of the use case.

Detailed Process:

- **Model Selection**: Depending on the type of data and task, select a suitable model architecture:
  - **For Text Generation**: Transformer-based models like **GPT-3**, **BERT**, or **T5** are effective.
  - **For Image Generation**: Consider models like **DALL-E**, **StyleGAN**, or **BigGAN**.
  - **For Multimodal Use Cases**: Use architectures like **CLIP** or **Florence** that combine text and image inputs.
- **Architecture Design**: Customize the model architecture based on the use case. This involves configuring the number of layers, attention heads, hidden units, and other parameters to suit the specific requirements.
- **Hyperparameter Setup**: Define initial hyperparameters such as learning rate, batch size, optimizer type, and regularization techniques.

Example: For a document summarization task, selecting the **T5 model** (Text-to-Text Transfer Transformer) is ideal, as it is specifically designed for NLP tasks like summarization and translation.

Best Practices:

- **Leverage Pre-trained Models**: Use pre-trained models whenever possible to reduce training time and cost.
- **Experiment with Model Variants**: Try multiple variants of a model (e.g., T5-small vs. T5-large) to find the optimal trade-off between performance and computational requirements.
- **Consider Scalability**: Choose models that can be scaled for distributed training or adapted for different deployment environments.

1.3. Model Training

Objective: Train the selected model using the prepared data to optimize its parameters.

Detailed Process:

- **Data Splitting**: Split the dataset into training, validation, and test sets. For large-scale Generative AI models, ensure that the test set covers diverse scenarios.
- **Training Iterations**: Train the model on the training set, adjusting weights based on loss and other performance metrics.
- **Validation During Training**: Use the validation set to monitor performance and prevent overfitting.

Best Practices:

- **Use Early Stopping**: Stop training if the validation loss stops improving to avoid overfitting.
- **Monitor Resource Usage**: Optimize batch sizes and learning rates to make efficient use of GPU or TPU resources.
- **Log Training Metrics**: Use tools like TensorBoard for real-time tracking of metrics.

1.4. Evaluation and Validation

Objective: Assess the model's performance on unseen data to ensure its generalizability.

Detailed Process:

- **Performance Metrics**: Use appropriate metrics like BLEU, ROUGE, F1-score, and accuracy.
- **Error Analysis**: Manually review outputs to identify patterns of errors.

1.5. Fine-Tuning and Optimization

Objective: Enhance the model's performance through additional training on specialized data or by adjusting model parameters.

Detailed Process:

- **Transfer Learning**: Fine-tune a pre-trained model on domain-specific data to adapt it for a new, specialized use case. For example, start with a general-purpose language model like GPT-3 and fine-tune it on a dataset of legal documents to create a legal document summarization model.
- **Hyperparameter Tuning**: Optimize the model by experimenting with hyperparameters such as learning rate, batch size, number of training epochs, and dropout rates. Automated hyperparameter tuning can be done using tools like **AWS SageMaker Tuner**, **Google Cloud Vertex AI Vizier**, or **Azure HyperDrive**.
- **Regularization Techniques**: Apply techniques like dropout, weight decay, and gradient clipping to prevent overfitting and stabilize training.

Example: Fine-tune a BERT model on a small dataset of customer service conversations to create a specialized chatbot model that understands industry-specific terminology.

Best Practices:

- **Start with Smaller Learning Rates**: Use lower learning rates when fine-tuning large models to prevent drastic weight updates.
- **Use Automated Hyperparameter Tuning**: Implement automated tuning techniques to explore a wide range of hyperparameters efficiently.
- **Regularize for Stability**: Apply dropout and other regularization techniques during fine-tuning to prevent overfitting, especially with smaller datasets.

1.6. Model Deployment and Integration

Objective: Deploy the trained model to production and integrate it with the target application for real-time or batch inference.

Detailed Process:

- **Deployment Options**: Choose between real-time deployment (e.g., REST API endpoints) or batch deployment based on the use case. For example, a chatbot requires real-time deployment, whereas a document summarization tool can use batch deployment for generating summaries in bulk.
- **Scalable Deployment**: Use managed services like **AWS SageMaker Endpoints**, **Google Cloud AI Platform Prediction**, or **Azure Machine Learning Endpoints** for scalable and reliable deployment.
- **Integration with Applications**: Integrate the deployed model into the target application using APIs. Ensure compatibility with existing software stacks, databases, and front-end interfaces.

Example: Deploy a text generation model as a RESTful API and integrate it into a chatbot system, enabling real-time response generation based on user queries.

Best Practices:

- **Use Auto-Scaling**: Enable auto-scaling to handle varying levels of traffic without manual intervention.
- **Implement Security Measures**: Use authentication, authorization, and encryption to secure API endpoints.
- **Test in Multiple Environments**: Test the model in different environments (development, staging, and production) to ensure consistent performance.

1.7. Continuous Monitoring and Maintenance

Objective: Continuously monitor model performance, detect drift, and update the model to maintain optimal performance.

Detailed Process:

- Monitoring Model Performance: Use tools like Amazon SageMaker Model Monitor, Google Cloud Vertex AI Monitoring, or Azure Monitor to track metrics such as latency, throughput, and accuracy.
- Detecting Data and Concept Drift: Implement mechanisms to detect when the underlying data distribution changes, causing the model to become less effective (concept drift). If drift is detected, initiate a retraining or fine-tuning process.
- Logging and Alerting: Set up alerts for anomalies in model outputs or sudden changes in key metrics.

Example: Monitor a deployed text summarization model for concept drift, such as when new topics or terminologies become prominent in the content being summarized. If drift is detected, retrain the model with updated data.

Best Practices:

- **Automate Monitoring**: Implement automated monitoring pipelines to reduce manual intervention.
- **Use Shadow Testing**: Deploy new model versions alongside the current production model (shadow testing) to evaluate their performance before switching.
- **Plan for Regular Updates**: Schedule periodic retraining to incorporate new data and maintain model relevance.

2 Creation of a Large Language Model (LLM)

Large Language Models (LLMs) are deep learning models designed to understand, generate, and manipulate natural language. They are typically built using the transformer architecture, which leverages self-attention mechanisms to capture contextual relationships between words. LLMs are characterized by their scale (often containing billions of parameters) and their ability to perform a wide variety of language tasks, such as text generation, summarization, translation, and question answering.

Large Language Models (LLMs) such as GPT-3, BERT, and T5 have set a new benchmark in Natural Language Processing (NLP) by demonstrating unprecedented capabilities in understanding and generating human language. Building a successful LLM from scratch involves overcoming significant challenges related to data, computational power, and architecture complexity. It requires meticulous planning, specialized tools, and access to powerful hardware.

This section provides a detailed, step-by-step guide on how to create an LLM, covering everything from data collection to deployment. We'll explore the tools, data requirements, and computational resources needed at each stage, along with time estimates and best practices to ensure that the process is as efficient as possible.

Key Characteristics of LLMs:

- **Large Scale:** LLMs are built with billions of parameters, enabling them to capture complex patterns and nuances in language.
- **Contextual Understanding:** They can maintain context over long sequences, making them suitable for generating coherent paragraphs or handling multi-turn conversations.
- **Generative Capabilities:** LLMs are capable of creating high-quality text, answering complex questions, and even generating creative content such as poetry or stories.

Building an LLM involves multiple stages, including data collection, model design, training, and deployment. The amount of time, data, and computational resources required varies significantly based on the scale and complexity of the model.

The process of creating an LLM can be divided into the following stages:

1. Data Collection and Preparation
2. Model Design and Architecture Selection
3. Model Training
4. Evaluation and Validation
5. Fine-Tuning and Optimization
6. Deployment and Integration
7. Monitoring and Maintenance

Each stage requires careful planning and implementation, and each has specific requirements in terms of time, data, and compute power.

3. Detailed Process for Building a Large Language Model

3.1. Data Collection and Preparation

Objective: Gather, clean, and prepare high-quality data that is representative of the language or domain the model is intended to operate in.

Time Required: 1-3 months (depending on the scope and diversity of data sources)

Data Requirements:

- **Volume:** For a large-scale LLM, at least 100 GB to 1 TB of text data is required. For models like GPT-3, the dataset size can exceed 500 billion tokens.
- **Data Sources:** Text data can be sourced from books, academic papers, websites, and proprietary documents. For domain-specific models (e.g., legal or medical), domain-specific text corpora are needed.

Process:

1. **Data Sourcing:** Identify and acquire text data from diverse sources such as Wikipedia, Common Crawl, Project Gutenberg, or proprietary sources.
2. **Data Cleaning:** Implement data preprocessing steps to remove noise, correct errors, and standardize text formatting. For example, remove HTML tags, special characters, and irrelevant content.
3. **Data Transformation:** Tokenize the text data using tools like NLTK, SpaCy, or Hugging Face Tokenizers to convert sentences into sequences of tokens.
4. **Data Annotation (Optional):** If building a supervised LLM, label the data with categories, tags, or context-specific information.

Tools:

- **AWS Glue:** For large-scale data cleaning and transformation.
- **Google Cloud Dataprep:** For data profiling and preparation.
- **Hugging Face Tokenizers:** For efficient text tokenization and data preparation.

Compute Requirements: Standard CPUs and distributed data processing frameworks like Apache Spark are sufficient for this stage.

Example: To create a multilingual LLM, collect datasets from multilingual Wikipedia pages, news articles, and translated books. Clean and preprocess these datasets to ensure they represent a wide variety of language constructs.

Best Practices:

- **Ensure Data Diversity:** Include a mix of formal, informal, and technical texts to improve the model's versatility.
- **Automate Data Cleaning:** Use automated scripts for repetitive tasks to reduce manual errors.
- **Document Data Sources and Transformations:** Maintain a record of all data sources and preprocessing steps for reproducibility and compliance.

3.2. Model Design and Architecture Selection

Objective: Select the appropriate model architecture based on the use case, data size, and target application.

Time Required: 2-4 weeks (for architecture experimentation and hyperparameter tuning)

Architecture Options:

- **Transformers:** The core architecture for most modern LLMs.
    - **GPT-Style Models:** Suitable for generative tasks like text generation.
    - **BERT-Style Models:** Effective for understanding tasks such as text classification and entity recognition.
    - **T5 (Text-to-Text Transfer Transformer):** A flexible architecture that can handle various text-to-text tasks such as translation and summarization.

Design Considerations:

1. **Number of Layers:** Determines the depth and complexity of the model.
2. **Attention Heads:** Higher attention heads improve the model's ability to capture complex relationships.
3. **Model Size:** Choose between small, medium, and large variants based on the available computational resources.

Tools:

- **PyTorch and TensorFlow:** For implementing custom model architectures.
- **Hugging Face Transformers:** Provides pre-built transformer models and architecture configurations.

Compute Requirements: At least 16-32 GPUs for architecture experimentation and training small-scale models. For larger models, TPUs or multi-node GPU clusters are recommended.

Example: For a customer service chatbot, select a GPT-2 architecture with 12 layers and 12 attention heads to balance complexity and computational cost.

Best Practices:

- **Use Pre-Trained Models:** Fine-tuning a pre-trained model is more cost-effective and faster than training from scratch.
- **Experiment with Model Variants:** Try different variants (e.g., BERT-base vs. BERT-large) to identify the best architecture for the use case.

3.3. Model Training

Objective: Train the selected model architecture on the prepared dataset to optimize the model's parameters.

Time Required: 1-6 months (depending on model size and computational resources)

Data Requirements:

- **High-Volume Text Data:** For large LLMs like GPT-3, at least 500 billion tokens are required.

Training Strategies:

- **Distributed Training:** Use parallel training on multiple GPUs/TPUs to reduce time and handle large-scale models.
- **Gradient Accumulation:** Useful when the batch size is limited due to GPU memory constraints.

Compute Requirements:

- **GPUs/TPUs:** At least 32-64 GPUs or TPU Pods for large-scale models.
- **Storage:** High-speed storage (SSD) for handling large datasets and model checkpoints.

Example: Train a BERT model on 500 GB of text data using a multi-node GPU cluster with 64 NVIDIA V100 GPUs.

Best Practices:

- **Use Mixed Precision Training:** Reduces memory usage and speeds up training.
- **Implement Early Stopping:** Stop training when the validation loss stops improving.
- **Log Training Metrics:** Use tools like TensorBoard to monitor loss, accuracy, and other metrics.

3.4. Evaluation and Validation

Objective: Measure the performance and generalization ability of the trained model using various evaluation metrics. This stage ensures that the model meets the desired performance criteria and does not overfit the training data.

Time Required: 2-4 weeks (depending on the size of the validation and test datasets)

Evaluation Criteria:

- **Accuracy:** Measures the percentage of correct predictions.
- **F1-Score:** A balanced metric that considers both precision and recall, making it useful for imbalanced datasets.
- **Perplexity:** A measure of how well a probabilistic model predicts a sample. Lower perplexity indicates better predictive performance.
- **BLEU (Bilingual Evaluation Understudy):** Measures the similarity between machine-generated text and a reference text, used primarily for translation tasks.

- **ROUGE (Recall-Oriented Understudy for Gisting Evaluation):** Used for evaluating text summarization models, comparing the overlap of n-grams between the generated and reference summaries.

Process:

1. **Create a Test Set:** Use a separate test dataset that was not seen during training to evaluate the model's generalization ability.
2. **Run Inference:** Generate predictions on the test set and compute performance metrics.
3. **Error Analysis:** Analyze incorrect predictions to identify patterns and areas for improvement.

Tools:

- **Hugging Face Evaluate:** For standardized NLP metrics such as BLEU, ROUGE, and perplexity.
- **AWS SageMaker Clarify:** For detecting and explaining biases in model predictions.
- **Google Cloud AI Explanations:** For providing insights into model behavior.

Compute Requirements: Standard GPUs are typically sufficient for evaluation, but for large-scale models, a multi-GPU setup may be necessary to speed up the process.

Example: Evaluate a text summarization model on 10,000 test documents using ROUGE score. Conduct error analysis to determine if the model struggles with certain document lengths or topics.

Best Practices:

- **Use Multiple Metrics:** Evaluate models using a combination of metrics (e.g., BLEU, ROUGE, and perplexity) to get a holistic view of performance.
- **Perform Error Analysis:** Manually review a subset of errors to identify systematic issues in the model.
- **Set Performance Baselines:** Compare the model against simple baselines (e.g., random predictions) to ensure meaningful improvements.

3.5. Fine-Tuning and Optimization

Objective: Enhance the model's performance by further training on specialized data or optimizing hyperparameters.

Time Required: 1-3 months (depending on the complexity of the fine-tuning tasks)

Fine-Tuning Strategies:

- **Transfer Learning:** Fine-tune a general-purpose LLM on a smaller, task-specific dataset (e.g., fine-tuning BERT for sentiment analysis in the financial domain).
- **Domain Adaptation:** Adapt the model to a new domain by training on domain-specific data (e.g., legal, medical, or technical texts).
- **Hyperparameter Optimization:** Use tools like AWS SageMaker Tuner, Google Vertex AI Vizier, or Azure HyperDrive to find the optimal set of hyperparameters.

Process:

1. **Identify the Target Task:** Define the specific use case (e.g., question answering, text classification).
2. **Prepare the Fine-Tuning Dataset:** Create a labeled dataset with task-specific annotations.
3. **Train with Modified Hyperparameters:** Adjust learning rate, batch size, and optimizer settings for the fine-tuning process.
4. **Validate the Model:** Use a validation set to monitor improvements in task-specific metrics.

Tools:

- **Hugging Face Transformers:** For implementing and fine-tuning transformer models.
- **Ray Tune:** For distributed hyperparameter optimization.
- **Google Cloud AutoML:** For automated model optimization.

Compute Requirements: Similar to initial training, fine-tuning typically requires 16-32 GPUs or TPUs for large models. Smaller models can be fine-tuned on fewer resources.

Example: Fine-tune a GPT-3 model for customer service by using a dataset of customer interactions and their corresponding support resolutions. Optimize hyperparameters like learning rate and batch size to improve response generation quality.

Best Practices:

- **Use a Smaller Learning Rate:** Fine-tuning should start with a lower learning rate (e.g., 0.00001) to prevent catastrophic forgetting.
- **Perform Regular Evaluation:** Continuously evaluate the model on a validation set to avoid overfitting.
- **Document Changes:** Track all modifications to the model during fine-tuning for reproducibility.

3.6. Deployment and Integration

Objective: Deploy the LLM to a production environment and integrate it with the target application for real-time or batch inference.

Time Required: 2-4 weeks (for integration, testing, and scaling)

Deployment Strategies:

- Real-Time Inference: Set up RESTful API endpoints to serve predictions in real time. This is ideal for chatbots, virtual assistants, and interactive applications.
- Batch Inference: Use batch processing for use cases like document summarization, where predictions can be generated in bulk and do not require immediate responses.
- Edge Deployment: Deploy smaller, distilled versions of the model to edge devices for applications with latency constraints.

Process:

1. **Select a Deployment Platform:** Choose between cloud-based deployment (e.g., AWS, Google Cloud, Azure) and on-premises deployment.
2. **Set Up the Model Endpoint:** Use managed services like AWS SageMaker Endpoints, Google Cloud AI Platform, or Azure Machine Learning Endpoints.

3. **Integrate with the Application:** Use APIs to connect the model endpoint to the target application (e.g., a chatbot UI or a document processing pipeline).
4. **Implement Monitoring:** Set up logging and monitoring to track usage, latency, and performance.

Tools:

- **AWS SageMaker:** For scalable, real-time deployment.
- **Google Cloud Vertex AI:** For deploying and monitoring machine learning models.
- **Azure Machine Learning:** For deploying and managing ML models in production.

Compute Requirements:

- **Scalable Cloud Instances:** Use auto-scaling clusters for real-time APIs.
- **Edge Hardware (Optional):** For edge deployment, use lightweight versions of the model optimized for devices like NVIDIA Jetson.

Example: Deploy a GPT-2 model as a REST API for real-time text generation in a chatbot application. Use auto-scaling on AWS SageMaker to handle peak loads and monitor latency using CloudWatch.

Best Practices:

- **Enable Auto-Scaling:** Configure auto-scaling to handle varying levels of traffic without manual intervention.
- **Implement Security Measures:** Secure APIs with authentication and encryption to prevent unauthorized access.
- **Test Across Environments:** Deploy in staging and development environments before pushing to production.

3.7. Monitoring and Maintenance

Objective: Continuously monitor the model in production, detect performance degradation, and update the model as needed.

Time Required: Ongoing (continuous monitoring and periodic updates)

Monitoring Aspects:

- **Performance Metrics:** Monitor latency, throughput, and error rates.
- **Data and Concept Drift:** Detect changes in the underlying data distribution that may cause the model to perform poorly over time.
- **Security and Compliance:** Regularly audit the model for compliance with regulations and ethical standards.

Tools:

- **AWS SageMaker Model Monitor:** For detecting data and concept drift.
- **Google Cloud Vertex AI Model Monitoring:** For real-time performance tracking.
- **Azure Monitor:** For logging and tracking model metrics.

Best Practices:

- **Automate Retraining:** Set up automated pipelines to retrain the model when significant drift is detected.
- **Shadow Testing:** Use shadow models to test updates in parallel with the production model.
- **Document Maintenance Procedures:** Maintain clear documentation of all monitoring and maintenance procedures for compliance and reproducibility.

4. Data, Compute Power, and Time Requirements

Creating a large-scale LLM from scratch is a resource-intensive endeavor. Below is an estimation of data, compute power, and time requirements based on model size:

Model Size	Parameters	Data Size	Training Time	Compute Power
Small	100M	10 GB	2-4 weeks	8-16 GPUs
Medium	1B	100 GB	1-3 months	16-32 GPUs
Large (e.g., GPT-3)	175B	500+ GB	3-6 months	64+ GPUs or TPUs, Distributed Training on Multi-Nodes

Building a Large Language Model is a complex, resource-intensive process that requires meticulous planning and careful management of data, compute resources, and time. By following a structured approach and leveraging cloud-based tools, organizations can create high-quality LLMs tailored to their specific needs, driving innovation and enhancing business outcomes.

5. Choosing the Right LLM for Different Use Cases

Generative AI models come in various architectures, each suited for specific applications. Selecting the right model is crucial for optimizing performance and resource utilization. Below is a guide to choosing the appropriate LLM based on different use cases:

5.1. Text Generation and Summarization

- **Recommended Models:** GPT-3, T5, and BERT
- **Use Cases:** Blog post generation, automatic summarization, chatbot responses.
- **Best Practices:**
  - Fine-tune on domain-specific data to ensure relevance and context.
  - Use attention mechanisms like those in T5 for high-quality text summarization.

5.2. Conversational AI and Chatbots

- **Recommended Models:** DialoGPT, GPT-3, BERT
- **Use Cases:** Virtual assistants, customer support bots.
- **Best Practices:**
  - Implement context handling techniques to maintain coherence across conversations.
  - Use reinforcement learning from human feedback (RLHF) to optimize responses.

5.3. Image and Video Generation

- **Recommended Models:** DALL-E, BigGAN, StyleGAN
- **Use Cases:** Image synthesis, style transfer, video generation.
- **Best Practices:**
  - Use high-resolution training datasets for better image quality.
  - Fine-tune using specific image styles or art forms to control output aesthetics.

5.4. Multimodal Models

- **Recommended Models:** CLIP, Florence, ALIGN
- **Use Cases:** Image captioning, cross-modal retrieval, text-to-image generation.
- **Best Practices:**
  - Pre-train on large-scale, diverse datasets to ensure robust cross-modal understanding.
  - Use contrastive learning techniques to improve alignment between modalities.

6. Conclusion

The Generative AI end-to-end process flow involves multiple stages, each playing a crucial role in building, deploying, and maintaining high-quality models. By carefully planning each step—data preparation, model selection, training, deployment, and continuous monitoring—organizations can maximize the impact of their Generative AI projects. Additionally, choosing the right LLM for specific use cases ensures that models are optimized for performance, cost, and scalability.

This chapter provides a comprehensive guide to navigating the complexities of Generative AI, equipping readers with the tools and best practices needed to successfully implement and manage these advanced AI systems.

Chapter 3

Architecture Options for Creating a Large Language Model (LLM)

As Large Language Models (LLMs) continue to revolutionize the fields of Natural Language Processing (NLP) and Artificial Intelligence, choosing the right architecture has become a pivotal factor in determining the performance, efficiency, and capabilities of these models. The foundational architectures, such as the Standard Transformer, have paved the way for a new generation of models like BERT, GPT, and T5. Meanwhile, newer architectures like Sparse Transformers, Longformer, Reformer, and Performer address the limitations of traditional models by optimizing for longer contexts, efficiency, and scalability.

In this chapter, we delve into various architecture options for building LLMs, from classic transformers to cutting-edge sparse and multimodal models, covering their design principles, use cases, strengths, limitations, and best practices. Understanding these architectures is crucial for selecting the right model for your specific use case, ensuring that your LLM is optimized for performance, scalability, and task relevance.

1. Overview of LLM Architectures

The architecture of an LLM defines how the model learns patterns, captures context, and generates outputs. Each architecture is designed with specific goals, whether it's reducing computation, handling long sequences, or integrating multimodal data. The most prominent architecture types include:

- Standard Transformer: The foundation for modern LLMs.
- GPT (Generative Pretrained Transformer): A decoder-only model for text generation.
- BERT (Bidirectional Encoder Representations from Transformers): An encoder-only model for contextual understanding.
- T5 (Text-to-Text Transfer Transformer): A versatile encoder-decoder model for text-to-text tasks.
- Sparse Transformers: Efficient models for long sequences.

- Longformer: A model for handling long-context documents.
- Reformer: A memory-efficient architecture for large sequences.
- Performer: A linear-time transformer using kernel-based attention.
- Routing Transformers: Dynamic attention models for selective attention.
- Multimodal Models (CLIP, Florence): Integrating text, image, and other modalities.

Each architecture has unique components, such as self-attention mechanisms, masking strategies, or sparse attention patterns, designed to optimize for different tasks and data characteristics.

2. Classic LLM Architectures

2.1. Standard Transformer (Vanilla Transformer)

The Standard Transformer architecture, introduced in the paper "Attention is All You Need," forms the foundation of almost all modern LLMs. It relies on self-attention mechanisms to capture dependencies across sequences, making it highly effective for various text-based tasks.

Architecture Design:

Encoder Stack: Composed of multiple layers, each with a Multi-Head Self-Attention mechanism and a Feed-Forward Network. The encoder captures the contextual representation of the input.

Decoder Stack: Also consists of multiple layers, each with a Masked Self-Attention, Encoder-Decoder Attention, and Feed-Forward Network. The decoder generates output sequences based on the encoder representation.

Use Cases:

- Suitable for tasks requiring both input and output sequences, such as translation and summarization.
- Forms the backbone for more specialized architectures like BERT and T5.

Strengths:

- Captures long-range dependencies effectively.
- Highly parallelizable due to the absence of recurrence.
- Limitations:
- High computational cost for long sequences due to O(n2)O(n^2)O(n2) complexity.

Visual Architecture Diagram: The diagram would show the encoder and decoder stacks connected through cross-attention layers, with arrows indicating data flow between layers.

2.2. GPT (Generative Pretrained Transformer)

GPT is a decoder-only transformer designed for generative tasks. It predicts the next token in a sequence using unidirectional attention, making it ideal for text generation, code completion, and dialogue systems.

Architecture Design:

Decoder-Only Stack: The model consists of multiple decoder layers, each with masked self-attention to prevent the model from seeing future tokens during training.

Causal Language Modeling: Trained to predict the next word given the previous context.

Use Cases:

- Text generation (e.g., story writing, chatbots).
- Few-shot learning (e.g., generating structured outputs from minimal examples).

Strengths:

- Excels at generating long-form text that maintains coherence.
- Can handle creative tasks like code generation (e.g., Codex).

Limitations:

- Limited to left-to-right context, making it less effective for understanding tasks.

Visual Architecture Diagram: A series of decoder layers with arrows representing left-to-right data flow and masked attention blocks.

2.3. BERT (Bidirectional Encoder Representations from Transformers)

BERT is an encoder-only transformer model designed to capture context in both directions. It is trained using Masked Language Modeling and Next Sentence Prediction, making it effective for understanding tasks.

Architecture Design:

Bidirectional Attention: Reads the entire sequence at once, using attention to capture dependencies in both directions.

Masked Language Modeling: Randomly masks words in the input and trains the model to predict them.

Use Cases:

- Text classification (e.g., sentiment analysis).
- Named entity recognition (NER).
- Question answering.

Strengths:

- Rich contextual embeddings.
- Strong performance on understanding tasks.

Limitations:

- Not suitable for text generation due to lack of a decoder.

Visual Architecture Diagram: Multiple encoder layers with arrows indicating bidirectional attention.

2.4. T5 (Text-to-Text Transfer Transformer)

T5 reframes all NLP tasks as text-to-text problems, using a unified encoder-decoder architecture. This enables T5 to handle a wide variety of tasks, from translation to summarization and classification, using a single model.

Architecture Design:

Encoder-Decoder Architecture: Converts all tasks into a text-to-text format, using both an encoder and a decoder stack.

Task-Specific Prompts: Each task is represented as a natural language prompt (e.g., "translate English to French:").

Use Cases:

- Summarization, question answering, translation.

Strengths:

- Unified format simplifies multi-task learning.
- Handles both generative and understanding tasks.

Limitations:

- High computational cost.

Visual Architecture Diagram: Show the encoder and decoder stacks with examples of text-to-text prompts.

3. New and Emerging Architectures

3.1. Sparse Transformers

Sparse Transformers reduce the quadratic complexity of standard transformers by introducing sparse attention patterns. By attending only to a subset of tokens, they achieve O(nlogn)O(n \log n)O(nlogn) complexity.

Use Cases: Long document summarization, handling large image sequences.

Strengths: Efficient for long sequences.

Limitations: May overlook critical dependencies if sparsity patterns are not well-designed.

Visual Architecture Diagram: Show sparse connections within the attention matrix.

3.2. Longformer

Longformer uses a combination of local and global attention to handle documents with up to 16,000 tokens.

Use Cases: Multi-paragraph QA, document classification.

Strengths: Efficiently processes long documents.

Limitations: Requires careful tuning of local and global attention ratios.

Visual Architecture Diagram: Illustrate a sliding window attention pattern with global tokens.

3.3. Reformer

Reformer introduces Locally Sensitive Hashing (LSH) and Reversible Layers to reduce the memory footprint of transformers.

Use Cases: Compressing long sequences, training large models.

Strengths: Memory-efficient and scalable.

Limitations: Complex implementation.

Visual Architecture Diagram: Show LSH blocks and reversible layers.

3.4. Performer

Performer uses kernel-based approximations to reduce the complexity of self-attention from O(n2)O(n^2)O(n2) to O(n)O(n)O(n).

Use Cases: Real-time inference, speech processing.

Strengths: Linear complexity, low latency.

Limitations: Approximation errors.

Visual Architecture Diagram: Show kernelized attention blocks.

3.5. Routing Transformers

Description: Routing Transformers are designed to dynamically adjust the attention patterns during training based on the content of the input sequence. Unlike standard transformers, which use static attention patterns, Routing Transformers compute attention selectively, making them more efficient and capable of capturing complex relationships in long sequences.

3.5.1. Architecture Design

Routing Transformers use content-based sparse attention, which allows the model to decide which tokens to attend to based on their content. The architecture includes:

Dynamic Attention Routing: Each token is assigned to a cluster based on its content, and attention is computed only between tokens in the same cluster. This is achieved using content-based token routing algorithms, similar to mixture-of-experts models.

Sparse Attention Matrix: Instead of attending to every token in a sequence, the model creates a sparse attention pattern, reducing computation.

3.5.2. Key Components

Content-Based Routing: Uses token similarities to dynamically route tokens into different attention clusters.

Sparse Attention Mechanism: Applies attention selectively based on the routing results.

3.5.3. Use Cases

- Dialogue Systems: Can focus on relevant parts of a conversation without attending to the entire history.
- Document Understanding: Suitable for handling document-level tasks that require selective attention to specific sections.

3.5.4. Strengths and Limitations

Strengths:

- Dynamic Attention: Adjusts attention based on content, capturing complex dependencies.
- Scalability: Efficiently handles longer sequences.

Limitations:

- Complexity: The dynamic routing mechanism can make the model harder to train.
- Sparsity Challenges: Requires careful tuning to ensure that all important tokens are attended to.

3.5.5. Visual Architecture Diagram

The diagram would include dynamic clusters for tokens, with arrows indicating content-based routing paths, and highlight the selective attention patterns formed between the token clusters.

4. Advanced Multimodal Models: CLIP, Florence, and Beyond

As AI systems evolve to handle more complex real-world scenarios, integrating multiple modalities (e.g., text, images, and audio) has become a key focus area. Multimodal models like CLIP and Florence represent a new generation of architectures designed to process and understand diverse data types in a unified manner.

4.1. CLIP (Contrastive Language-Image Pretraining)

Developer: OpenAI

Description: CLIP is a multimodal model trained to align visual and textual representations using a contrastive learning objective. It jointly trains a text transformer and a vision transformer to project both image and text data into a shared embedding space, making it capable of performing tasks such as image classification and zero-shot learning.

4.1.1. Architecture Design

CLIP consists of two separate components:

Vision Transformer (ViT):

Takes an image as input, divides it into patches, and processes these patches using a transformer-based encoder.

Produces a sequence of visual embeddings that capture spatial and semantic information.

Text Transformer:

Takes a text input, tokenizes it, and processes it using a standard transformer encoder.

Produces a sequence of word embeddings and a final contextual representation.

Shared Embedding Space:

Both the vision and text transformers are trained to project their outputs into a shared embedding space, where similar images and texts have high cosine similarity.

4.1.2. Training Paradigm

CLIP uses contrastive learning to maximize the similarity between correctly paired image-text inputs and minimize the similarity for incorrect pairs. The objective function is defined as:

$$\text{Loss} = -\frac{1}{N}\sum_{i=1}^{N}\log\frac{\exp(\text{sim}(I_i, T_i)/\tau)}{\sum_{j=1}^{N}\exp(\text{sim}(I_i, T_j)/\tau)}$$

sim(Ii,Ti) denotes the similarity between image Ii and text Ti, and τ is a temperature parameter.

4.1.3. Use Cases

- Image Classification: Performs classification based on natural language prompts.
- Image-Text Matching: Finds the best-matching image for a given text description.
- Zero-Shot Classification: Performs new classification tasks using only natural language prompts.

4.1.4. Strengths and Limitations

Strengths:

- Zero-Shot Capabilities: Can generalize to new tasks without additional training.
- Cross-Modal Understanding: Captures rich relationships between images and text.

Limitations:

Training Bias: Prone to biases present in the internet-based training data.

Resource Intensive: Requires significant computational resources for pretraining.

4.1.5. Visual Architecture Diagram

The diagram would show two separate transformers—one for vision and one for text—feeding into a shared embedding space, with arrows indicating cross-modal alignment.

4.2. Florence: Microsoft's Unified Vision-Language Model

Developer: Microsoft

Description: Florence extends the capabilities of multimodal models by incorporating Hierarchical Vision Transformers (HVT) and advanced cross-modal fusion techniques. It is designed to handle complex vision-language tasks, such as visual question answering, image captioning, and multimodal retrieval.

4.2.1. Architecture Design

Florence uses a dual-stream transformer design:

- Hierarchical Vision Transformer (HVT):
  - Uses a multi-scale approach to capture spatial hierarchies and fine-grained details in images.
  - Generates dense visual embeddings for different regions of the image.
- Language Transformer:
  - A traditional transformer-based text encoder that captures semantic information from text inputs.
- Cross-Modal Fusion Module:
  - Integrates the visual and textual embeddings at multiple layers using a cross-modal attention mechanism.

4.2.2. Training Paradigm

Florence is trained on large-scale multimodal datasets, using multi-task learning to perform both image classification and text-based tasks simultaneously. This enables it to generalize across different vision-language tasks.

4.2.3. Use Cases

- Visual Question Answering (VQA): Answers questions about images using natural language.
- Multimodal Retrieval: Finds the best-matching image for a given text or vice versa.
- Image Captioning: Generates detailed captions based on image content.

4.2.4. Strengths and Limitations

Strengths:

- Multi-Task Learning: Handles a variety of vision-language tasks without task-specific fine-tuning.
- Hierarchical Understanding: Captures both local and global visual information.

Limitations:

- High Computational Cost: Requires large-scale hardware for effective training.
- Complex Training Pipeline: Involves multi-stage training, making it challenging to implement.

4.2.5. Visual Architecture Diagram

The diagram would show separate streams for vision and text, with a cross-modal fusion module connecting the two, and examples of image-captioning outputs.

5. Conclusion

The architecture of an LLM is a key determinant of its capabilities, efficiency, and applicability. From classic models like BERT and GPT to advanced architectures like Sparse Transformers, Longformers, and multimodal models such as CLIP and Florence, understanding the strengths and limitations of each model is essential for building high-performing AI systems.

Selecting the right architecture requires careful consideration of the task, data, and computational resources. By leveraging these architectures and their innovations, organizations can build more efficient, scalable, and context-aware models, driving the next wave of AI advancements.

Chapter 4

Approach to Choosing LLM Models for Generative AI Use Cases

Generative AI is transforming industries by enabling automation in text generation, code completion, and content creation. The backbone of these capabilities lies in the underlying Large Language Models (LLMs). As organizations look to integrate Generative AI into their systems, choosing the right LLM becomes a strategic decision. This chapter provides a comprehensive approach to selecting LLMs, detailing each step in the decision-making process and providing insights into the various models available, their strengths and weaknesses, and the considerations necessary for deploying them effectively. The chapter concludes with best practices to ensure optimal model performance and alignment with business objectives.

1. Understanding Large Language Models (LLMs)

Large Language Models (LLMs) are deep learning models trained on vast amounts of text data, allowing them to understand, generate, and manipulate human language. Their ability to perform a wide range of Natural Language Processing (NLP) tasks makes them valuable across industries. Here, we explore the characteristics and features that define modern LLMs.

1.1. Key Features of LLMs

- **Text Understanding and Generation:**
  - LLMs like GPT-4 and PaLM 2 are capable of generating human-like text based on given prompts, making them ideal for content creation, summarization, and dialogue generation.
  - They can understand complex syntactic and semantic structures, allowing them to comprehend and respond to nuanced text inputs.

- **Contextual Awareness:**
    - Modern LLMs are equipped with mechanisms like transformers that enable them to retain context over long sequences. For example, models like GPT-4 can maintain a coherent conversation over multiple turns or generate lengthy documents that remain contextually relevant.
- **Knowledge Integration:**
    - Some LLMs are trained on vast datasets that include encyclopedic knowledge, making them useful for factual question answering and reasoning tasks. Models like GPT-4, T5, and PaLM 2 can leverage this internal knowledge to provide detailed, contextually accurate responses.
- **Multilingual Proficiency:**
    - Many LLMs, including PaLM and T5, are trained on multilingual corpora, enabling them to perform cross-lingual text generation and translation.
- **Adaptability:**
    - LLMs can be fine-tuned on domain-specific data to adapt to specialized tasks, such as legal document analysis, medical report generation, or technical writing.

1.2. Typical Use Cases for LLMs in Generative AI

- **Text Summarization:**
    - Use LLMs to create concise summaries of lengthy documents, research papers, or articles. For instance, fine-tuned BERT or GPT-3 models can condense legal contracts or scientific papers into digestible summaries.
- **Content Creation:**
    - LLMs are widely used for generating marketing copy, product descriptions, blogs, and even creative writing. Tools like OpenAI's GPT-4 can produce high-quality content with minimal human intervention.
- **Code Generation:**
    - LLMs like OpenAI's Codex are specifically designed for generating and completing code snippets, automating repetitive coding tasks, and assisting in software development.
- **Conversational AI:**
    - LLMs can power chatbots and virtual assistants, enabling them to hold natural conversations, answer queries, and provide contextual assistance.
- **Data Augmentation:**
    - LLMs can generate synthetic datasets for training other machine learning models, especially in scenarios where obtaining real-world data is challenging or expensive.
- **Knowledge-Based Reasoning:**
    - LLMs can be used to reason about complex problems, generate logical arguments, and provide structured outputs based on their understanding of a topic.

2. Approach to Selecting the Right LLM for Your Use Case

The process of choosing the right LLM involves evaluating multiple factors, such as model capabilities, infrastructure requirements, cost, and compliance. This section outlines a structured approach to help organizations make informed decisions.

2.1. Step 1: Define Use Case Requirements

Start by outlining the primary goals and requirements of the use case. Having a clear understanding of what you want the LLM to achieve will narrow down the list of potential models.

- **Primary Objective:**
  - What is the end goal? Are you looking to generate creative content, automate document summarization, or assist with coding tasks?
  - Example: For a legal firm, the objective might be to generate detailed summaries of case law or automate contract drafting.
- **Expected Outputs:**
  - Are you looking for long-form content, short responses, or structured data outputs? Understanding the nature of the output will influence the choice of model.
  - Example: If the output needs to be a structured FAQ, a model like BERT or T5 might be more suitable than GPT-3.
- **Level of Contextual Understanding:**
  - Does the model need to remember user inputs across multiple turns in a conversation? Models like GPT-4 are more suited for multi-turn dialogues, whereas others may be limited in their contextual awareness.
- **Language Requirements:**
  - If your use case requires multilingual support, consider models like PaLM or T5, which are optimized for cross-lingual tasks.
- **User Base:**
  - Define the user base (e.g., internal employees, customers) and the expected volume of interactions to determine scalability and performance needs.

2.2. Step 2: Evaluate Model Capabilities

Evaluate the LLMs based on their performance, architecture, and specific strengths. Each model has unique features that make it better suited for certain applications.

- **Performance and Accuracy:**
  - Use benchmark scores such as GLUE, SuperGLUE, and SQuAD to compare model performance on various NLP tasks. Choose models that have demonstrated high scores on benchmarks relevant to your use case.
- **Model Size and Complexity:**
  - Larger models tend to perform better but are computationally intensive. For example, GPT-3 has 175 billion parameters, making it suitable for complex text generation but resource-heavy.

- **Context Length:**
  - Models like GPT-4 can handle context lengths of up to 8,000 tokens, making them ideal for applications that require long-term memory, such as legal document analysis or academic research.
- **Domain Expertise:**
  - Some models are pre-trained or fine-tuned for specific domains, such as BioBERT for biomedical text. Choosing a domain-specific model can significantly improve accuracy for specialized tasks.
- **Latency and Throughput:**
  - Consider latency requirements for real-time applications. High-latency models may not be suitable for interactive systems like chatbots or customer support.

2.3. Step 3: Consider Operational Constraints

Operational considerations such as cost, deployment infrastructure, and scalability should influence the choice of LLM.

- **Cost of Inference:**
  - Larger models are more expensive to run. Evaluate the total cost of ownership, including the cost per API call or the expense of running the model on GPUs or TPUs.
- **Hardware and Scalability:**
  - Determine if your infrastructure can support the chosen model. Some models require specialized hardware like TPUs (e.g., Google's PaLM), while others can run on more standard hardware configurations.
- **Integration with Existing Systems:**
  - Ensure that the model can be seamlessly integrated into existing workflows, APIs, and cloud infrastructure.
- **Scalability:**
  - If the application involves high volumes of traffic, choose a model that can scale horizontally across multiple instances.

2.4. Step 4: Choose Between Pre-Trained and Fine-Tuned Models

Decide whether to use a general-purpose pre-trained model or a fine-tuned model specifically adapted for your use case.

- **Pre-Trained Models:**
  - These are suitable for generic applications, such as customer support or basic content generation. Models like GPT-4 and PaLM offer high versatility out-of-the-box.
- **Fine-Tuned Models:**
  - Fine-tuning involves training a pre-existing model on domain-specific data to improve its performance for specialized tasks. For example, fine-tuning BERT on a legal text corpus would result in a model highly adept at legal text interpretation.
- **Custom Models:**
  - For niche applications, consider training a custom model using frameworks like TensorFlow or PyTorch and deploying it on platforms like Vertex AI or SageMaker.

2.5. Step 5: Evaluate Compliance and Ethical Considerations

Large Language Models (LLMs) can pose risks related to data privacy, compliance, and ethical use. Ensuring that the chosen LLM meets industry standards and adheres to responsible AI principles is critical for minimizing risks and maintaining trust.

- **Data Privacy:**
  - LLMs trained on sensitive data may inadvertently memorize and expose private information. For applications involving personal data (e.g., customer records or healthcare information), choose models that have robust data handling policies.
  - **Best Practice**: Use fine-tuning and training methods that prevent data memorization, such as differential privacy techniques that add noise to sensitive inputs.
- **Bias and Fairness:**
  - LLMs can inherit biases present in their training data, leading to outputs that reflect societal, gender, or racial biases. Assess each model for bias using tools like Google's *What-If Tool* or OpenAI's *Bias and Fairness Evaluator.*
  - **Best Practice**: Regularly test models for biased outputs, especially in sensitive applications like hiring, lending, or customer support, and mitigate biases through fine-tuning and reinforcement learning.
- **Compliance:**
  - Consider industry-specific regulations (e.g., GDPR, HIPAA, CCPA) when selecting models. For example, models used in healthcare must comply with HIPAA regulations regarding the handling of patient data.
  - **Best Practice**: Use data masking, encryption, and access control mechanisms to ensure compliance.
- **Ethical Use:**
  - Avoid using LLMs for potentially harmful applications, such as generating misinformation or producing offensive content. Employ monitoring tools to detect and prevent misuse.
  - **Best Practice**: Implement usage policies that restrict the model from generating harmful or unethical content and apply filters to flag inappropriate responses.

By addressing these considerations, organizations can deploy LLMs responsibly and ensure that their Generative AI solutions align with ethical standards and regulatory requirements.

3. Major LLM Providers and Their Models

With the growing popularity of Generative AI, several companies now offer powerful LLMs tailored for different applications. The following is an overview of the major LLM providers, highlighting their flagship models, key features, and areas of specialization.

3.1. OpenAI

OpenAI is a leading provider of LLMs, offering a range of models known for their versatility and high performance in natural language understanding and generation tasks.

- **GPT-4:**
  - **Description**: OpenAI's most advanced model, capable of producing coherent, contextually aware text across a variety of domains.
  - **Key Features**: High accuracy, large context window, and support for complex reasoning tasks.
  - **Use Cases**: Content creation, conversational AI, legal and financial text analysis.
  - **Deployment Options**: Available through OpenAI's API and Azure OpenAI Service.
- **GPT-3.5:**
  - **Description**: A smaller and more cost-effective version of GPT-4, offering high performance for general NLP tasks.
  - **Key Features**: Multilingual support, optimized for efficiency.
  - **Use Cases**: Text summarization, chatbots, code generation.
  - **Deployment Options**: OpenAI API and Azure OpenAI Service.
- **Codex:**
  - **Description**: A model fine-tuned specifically for code generation and programming assistance.
  - **Key Features**: Supports over a dozen programming languages, including Python, JavaScript, and C++.
  - **Use Cases**: Code autocompletion, bug fixing, automation of repetitive coding tasks.
  - **Deployment Options**: Available through the OpenAI Codex API.

3.2. Google Cloud

Google Cloud's suite of LLMs is known for its integration with Google's powerful cloud services, making it a top choice for enterprises seeking scalable and reliable AI solutions.

- **PaLM 2:**
  - **Description**: Google's latest large language model, designed for high performance in multilingual tasks and complex reasoning.
  - **Key Features**: Advanced support for long context, improved factuality, and reasoning capabilities.
  - **Use Cases**: Legal document analysis, academic research, multilingual content generation.
  - **Deployment Options**: Available through Google Cloud's Vertex AI platform.
- **BERT:**
  - **Description**: A bi-directional transformer model optimized for tasks that require deep language understanding.
  - **Key Features**: Strong performance in sentiment analysis, question answering, and entity recognition.
  - **Use Cases**: Search engines, customer sentiment analysis, question answering.
  - **Deployment Options**: Available through TensorFlow Hub and Vertex AI.
- **T5 (Text-to-Text Transfer Transformer):**
  - **Description**: A versatile model that treats every NLP problem as a text-to-text problem, enabling unified handling of a variety of tasks.
  - **Key Features**: Strong performance across text classification, translation, summarization, and Q&A.
  - **Use Cases**: Text generation, translation, summarization.
  - **Deployment Options**: Available through TensorFlow Hub and Google Cloud APIs.

3.3. Microsoft Azure

Microsoft provides LLMs through its Azure OpenAI Service, making it easier for enterprises to integrate powerful language models into their existing Azure ecosystems.

- **Azure OpenAI Service:**
  - **Description**: Provides access to OpenAI's models, including GPT-4, GPT-3.5, and Codex, with Azure's enterprise-grade security and compliance.
  - **Key Features**: Integrated with Azure security and compliance tools, making it suitable for regulated industries.
  - **Use Cases**: Enterprise chatbot development, document automation, and financial analysis.
  - **Deployment Options**: Available through Azure OpenAI Service.
- **Turing-NLG:**
  - **Description**: Microsoft's in-house language model, designed for high-quality text generation and understanding.
  - **Key Features**: Supports a variety of NLP tasks, including summarization and dialogue generation.
  - **Use Cases**: Virtual assistants, content creation, automated document review.
  - **Deployment Options**: Integrated into Azure Cognitive Services.

3.4. Anthropic

Anthropic is a new entrant focused on building models with safety and alignment at the forefront.

- **Claude:**
  - **Description**: A language model designed to be safer and more aligned with human values, reducing the risk of harmful outputs.
  - **Key Features**: Advanced safety mechanisms, reliable performance in text generation.
  - **Use Cases**: Customer support, educational tools, interactive learning.
  - **Deployment Options**: Available through Anthropic's API.

3.5. Cohere

Cohere offers models optimized for NLP tasks such as text classification, search, and retrieval.

- **Command R:**
  - **Description**: A retrieval-augmented generation model designed for high-quality, long-form content creation.
  - **Key Features**: Combines generative capabilities with information retrieval for enhanced factual accuracy.
  - **Use Cases**: Research article writing, technical documentation, blog generation.
  - **Deployment Options**: Available through Cohere's API.
- **Command-X:**
  - **Description**: A high-performance model optimized for efficiency and low-latency applications.
  - **Key Features**: Fast response times, suitable for real-time interactions.
  - **Use Cases**: Chatbots, real-time translation, question answering.

  - **Deployment Options**: Cohere's API.

3.6. AI21 Labs

AI21 Labs focuses on building models that offer a high degree of customization and flexibility for various NLP applications. Their models are known for their multilingual capabilities and fine-grained control over text generation.

- **Jurassic-2:**
  - **Description**: AI21's flagship LLM, designed for high performance across a broad range of language tasks.
  - **Key Features**: Strong multilingual capabilities, control over generation style and tone, fine-tuned variants for specific domains.
  - **Use Cases**: Content creation, technical writing, legal and financial text analysis.
  - **Deployment Options**: Available through the AI21 Studio API and integrated into AI21 Labs' commercial products.
- **Jurassic-2 Jumbo:**
  - **Description**: A larger variant of Jurassic-2, optimized for complex language tasks and detailed reasoning.
  - **Key Features**: Larger model size with enhanced accuracy and contextual understanding.
  - **Use Cases**: Long-form content generation, research document synthesis, and multilingual translation.
  - **Deployment Options**: AI21 Studio API.

3.7. Meta AI

Meta AI (formerly Facebook AI) offers open-source models focused on enabling research and development in natural language processing.

- **LLaMA (Large Language Model Meta AI):**
  - **Description**: A suite of highly efficient models designed to be accessible and resource-efficient while delivering strong performance.
  - **Key Features**: Open-source, optimized for research, supports a variety of tasks from text generation to question answering.
  - **Use Cases**: Academic research, experimental applications, and NLP training tasks.
  - **Deployment Options**: Available through Meta's GitHub repository and as open-source models.
- **LLaMA 2:**
  - **Description**: The next generation of LLaMA with improved performance, larger context windows, and support for fine-tuning.
  - **Key Features**: High efficiency, low latency, and suitable for deployment in research and enterprise environments.
  - **Use Cases**: Text summarization, dialogue systems, and multi-turn question answering.
  - **Deployment Options**: Available as open-source models.

3.8. Stability AI

Stability AI, primarily known for its work in generative art and image models (e.g., Stable Diffusion), has extended its research into text-based models with an emphasis on open-source solutions.

- **StableLM:**
  - **Description**: An open-source language model aimed at providing a transparent and community-driven alternative to proprietary LLMs.
  - **Key Features**: Lightweight, designed for community contributions and improvements, transparency in training data and methods.
  - **Use Cases**: Educational tools, research, and lightweight text generation.
  - **Deployment Options**: Available through Stability AI's repositories and compatible with various open-source platforms.
- **StableLM-X:**
  - **Description**: An advanced version of StableLM, optimized for specific applications like creative writing and dialogue generation.
  - **Key Features**: Higher contextual awareness, improved coherence in long-form text generation.
  - **Use Cases**: Creative content creation, collaborative writing, and educational chatbots.
  - **Deployment Options**: Open-source distribution and APIs.

4. Best Practices for Choosing an LLM

Selecting the right LLM is a nuanced process that involves understanding both the technical and business aspects of model deployment. The following best practices provide a structured approach to guide organizations in choosing the optimal LLM for their use cases.

4.1. Prioritize Domain-Specific Models When Necessary

For applications that operate within specialized domains (e.g., legal, finance, healthcare), consider using models that have been fine-tuned on domain-specific data. These models typically outperform general-purpose LLMs in tasks that require specialized knowledge or precise language use.

- **Example**: Use BioBERT for biomedical text analysis or FinBERT for financial sentiment analysis.
- **Consideration:**
  - **Accuracy vs. Flexibility**: Domain-specific models offer higher accuracy but may lack versatility outside their trained domain.
- **Best Practice**: Evaluate the performance of general-purpose models against domain-specific benchmarks to decide if the accuracy gain justifies the additional fine-tuning effort and cost.

4.2. Balance Model Size and Cost

While larger models (e.g., GPT-4, Jurassic-2 Jumbo) offer superior performance, they also come with significantly higher operational costs. For cost-sensitive applications, consider smaller, distilled versions of the models that strike a balance between performance and efficiency.

- **Example**: Opt for GPT-3.5 instead of GPT-4 for applications that do not require the highest level of accuracy.
- **Consideration**:
  - **Trade-Off**: Smaller models have lower latency and cost but may sacrifice contextual understanding and fine-grained text generation quality.
- **Best Practice**: Use cost estimators and performance benchmarks provided by LLM providers to determine the optimal model size for your budget.

4.3. Use Latency-Optimized Models for Real-Time Applications

For real-time applications such as chatbots, customer service, or interactive assistants, latency is a critical factor. Choose models that offer low response times and high throughput to ensure smooth user experiences.

- **Example**: Use Cohere's Command-X for real-time chat applications due to its latency-optimized architecture.
- **Consideration**:
  - **Model Architecture**: Some models are designed with latency optimization in mind, using techniques like reduced parameter sizes and faster attention mechanisms.
- **Best Practice**: Test multiple models in real-world scenarios to measure latency and throughput, and choose the one that offers the best performance within acceptable limits.

4.4. Perform Rigorous Evaluation and Testing

Before selecting an LLM for production use, conduct thorough testing using real-world data and relevant benchmarks. Measure performance across multiple dimensions, including accuracy, fluency, consistency, and bias.

- **Example**: Use test datasets that reflect the expected input types and formats in the production environment to evaluate how well the model generalizes.
- **Consideration**:
  - **Evaluation Metrics**: Use metrics like BLEU, ROUGE, and perplexity for text generation, and accuracy, F1-score, or AUC for classification tasks.
- **Best Practice**: Include human evaluation as part of the testing process, especially for applications that generate creative or subjective content.

4.5. Implement Robust Security and Compliance Measures

LLMs can process sensitive information, making security and compliance a top priority. Use techniques such as data encryption, access control, and regular audits to ensure that the deployment adheres to security standards and regulatory requirements.

- **Example**: Implement VPC Service Controls when deploying models on Google Cloud to restrict access to sensitive datasets and prevent data leakage.
- **Consideration**:
  - **Data Handling**: Choose models that support secure data handling policies, especially if processing personal or proprietary information.

- **Best Practice**: Use compliance frameworks (e.g., GDPR, HIPAA) as a checklist for evaluating the deployment strategy and ensure that all data interactions are logged and monitored.

4.6. Regularly Update and Fine-Tune Models

LLMs evolve rapidly, and new research often leads to significant improvements in performance, efficiency, and safety. Regularly updating models to incorporate these advancements can provide better outcomes and maintain competitiveness.

- **Example**: Transition from older models like GPT-3 to newer versions like GPT-4 or LLaMA 2 to take advantage of improved reasoning and factuality.

- **Consideration**:
  - **Model Drift**: Regularly update and revalidate fine-tuned models to prevent model drift and ensure alignment with current business objectives.
- **Best Practice**: Establish a model update schedule that aligns with product release cycles and business needs, and automate testing and validation processes to streamline upgrades.

5. Considerations When Choosing an LLM Provider

Selecting the right LLM provider is as crucial as choosing the model itself. Providers differ in terms of deployment options, pricing, support, and integration capabilities. Below are key considerations for evaluating LLM providers:

Integration Capabilities

- **Seamless Integration:**
  - Ensure that the LLM provider offers APIs and SDKs that are compatible with your existing tech stack (e.g., Python, RESTful APIs).
  - **Example**: Google's Vertex AI provides deep integration with Google Cloud services, making it easier to incorporate LLMs into existing GCP workflows.
- **Ecosystem Support:**
  - Look for providers that support integration with popular machine learning frameworks like TensorFlow, PyTorch, or Hugging Face.

Scalability and Performance

- **Compute Infrastructure:**
  - Choose a provider that offers the necessary compute resources (e.g., GPUs, TPUs) to scale model deployments without compromising performance.
  - **Example**: Google Cloud's TPU v4 Pods are optimized for training and deploying large-scale LLMs like PaLM.
- **Load Management:**
  - Ensure that the provider supports autoscaling and load balancing to handle fluctuations in demand.

Compliance and Security

- **Compliance Standards:**
  - Ensure that the provider complies with relevant industry standards and regulations such as GDPR, HIPAA, SOC 2, or PCI DSS. This is crucial for organizations in regulated industries such as healthcare, finance, and government.
  - **Example**: Microsoft Azure's OpenAI service integrates with Azure's compliance and security tools, making it suitable for enterprise-grade applications that require rigorous security controls.
- **Data Security and Privacy:**
  - Verify that the provider offers robust data protection mechanisms, including end-to-end encryption, access control, and secure storage. If the model processes sensitive information, ensure that it adheres to strict data privacy policies.
  - **Example**: Anthropic's Claude model is designed with safety and privacy at its core, offering options for secure data handling and minimization of sensitive information retention.
- **Service-Level Agreements (SLAs):**
  - Check the provider's SLAs for availability, reliability, and response times, especially for mission-critical applications. Providers like Google Cloud and Azure offer SLAs that guarantee high availability and fast support.

Support and Customization

- **Dedicated Support:**
  - Providers that offer dedicated support, including AI specialists and solution architects, can significantly reduce deployment time and ensure smooth integration.
  - **Example**: OpenAI's enterprise plans offer dedicated support, including access to OpenAI engineers and model customization guidance.
- **Customization and Fine-Tuning:**
  - Consider whether the provider supports fine-tuning or custom training on your proprietary data. Some providers, like Cohere, focus on providing easy-to-use fine-tuning tools for their models.
  - **Best Practice**: Opt for providers that offer built-in tools for training, evaluation, and optimization to create custom models suited to your unique requirements.

Pricing and Licensing Models

- **Pay-As-You-Go vs. Subscription:**
  - Evaluate different pricing models to choose the one that aligns with your usage patterns. Pay-as-you-go models are ideal for experimentation, while subscription models may be more cost-effective for large-scale deployments.
  - **Example**: OpenAI offers pay-per-use pricing based on token consumption, while providers like Cohere offer monthly subscriptions for higher-volume users.
- **License Flexibility:**
  - For open-source models like LLaMA or StableLM, consider the licensing agreements, especially if you plan to modify or commercialize the model. Some models have restrictive licenses that may limit commercial use.

6. Best Practices for Model Deployment and Operationalization

Deploying an LLM is not the end of the journey; maintaining its performance, ensuring security, and optimizing for cost efficiency are continuous processes. Below are best practices for deploying and operationalizing LLMs effectively.

6.1. Optimize for Cost and Resource Efficiency

LLMs, especially large ones, can be computationally expensive to run. Cost optimization strategies can help ensure that resources are used efficiently without compromising performance.

- **Use Model Pruning and Quantization:**
  - Techniques like pruning and quantization can reduce model size and inference time, lowering costs while maintaining acceptable performance.
  - **Example**: Quantizing a GPT-3 model to use 8-bit precision instead of 16-bit can reduce the memory footprint and computational cost.
- **Deploy Scaled-Down Variants for Less Intensive Tasks:**
  - Use smaller versions of the model for less demanding tasks. For example, a scaled-down variant of BERT (e.g., DistilBERT) can handle lightweight tasks like text classification with lower latency.
- **Leverage On-Demand and Spot Instances:**
  - Consider using spot instances or preemptible VMs for non-critical workloads to save costs. Spot instances offer significant discounts compared to on-demand VMs, making them ideal for batch processing.

6.2. Implement Security Best Practices for Deployment

Deploying LLMs requires stringent security controls to prevent unauthorized access, data leaks, and misuse.

- **Use VPCs and Network Isolation:**
  - Deploy models within a Virtual Private Cloud (VPC) to isolate them from the public internet. Use network firewalls and security groups to restrict access.
  - **Example**: On Google Cloud, Vertex AI models can be deployed within a VPC Service Control perimeter, ensuring that only trusted services and users can access the models.
- **Implement Role-Based Access Control (RBAC):**
  - Use RBAC to define roles and permissions for different users, such as data scientists, engineers, and business users. This minimizes the risk of unauthorized changes or data access.
  - **Best Practice**: Regularly audit access logs and permissions to ensure compliance and detect potential security issues.
- **Use API Gateways for Secure Access:**
  - Expose model endpoints using API gateways with built-in authentication and rate limiting. This prevents misuse and protects against distributed denial-of-service (DDoS) attacks.
  - **Example**: Use Google Cloud Endpoints or AWS API Gateway with IAM-based authentication to secure API access.

6.3. Implement Monitoring and Observability

Monitoring is essential to detect issues such as model drift, reduced performance, or security vulnerabilities in real-time.

- **Set Up Model Performance Monitoring:**
  - Use tools like Google Cloud's Vertex AI Model Monitoring or Amazon SageMaker Model Monitor to track model performance metrics, such as latency, error rates, and accuracy.
  - **Best Practice**: Set thresholds for key performance indicators (KPIs) and configure alerts to detect anomalies early.
- **Monitor for Model Drift and Bias:**
  - Implement monitoring to detect model drift—when a model's performance degrades due to changes in the input data distribution. Regularly test for bias and fairness to ensure the model remains aligned with ethical guidelines.
  - **Example**: Use drift detection tools like Vertex AI's built-in drift monitors or custom drift detection scripts using scikit-learn.
- **Log and Trace All Interactions:**
  - Enable logging and tracing of all API calls, inputs, and outputs to create an audit trail. This is crucial for compliance and troubleshooting.
  - **Example**: Use Cloud Logging on GCP or CloudWatch on AWS to centralize logs and create automated alerts for suspicious activities.

6.4. Regularly Update and Fine-Tune Models

Regular updates and fine-tuning are necessary to keep LLMs performing optimally as data, user needs, and business goals evolve.

- **Retrain Models on Updated Data:**
  - As new data becomes available, retrain or fine-tune models to incorporate the latest information. This is particularly important for domains with rapidly changing knowledge, such as finance or technology.
  - **Best Practice**: Implement automated retraining pipelines using MLOps frameworks like Kubeflow or TFX to streamline the retraining process.
- **Adapt Models to Changing User Behavior:**
  - Use reinforcement learning and active learning techniques to adapt models based on real-time feedback and user interactions.
  - **Example**: Deploy a reinforcement learning loop to update the model's response generation strategy based on user feedback.
- **Evaluate Models Against New Benchmarks:**
  - Continuously test models against updated benchmarks and real-world scenarios to identify areas for improvement.
  - **Best Practice**: Use standardized benchmarks (e.g., SuperGLUE, SQuAD) and internal performance tests to evaluate model updates.

7. Conclusion

Choosing the right LLM for a Generative AI use case requires careful consideration of the model's capabilities, operational requirements, compliance constraints, and business objectives. By following the structured approach and best practices outlined in this chapter, organizations can select models that align with their goals and deploy them effectively, ensuring optimal performance, security, and cost efficiency. With the rapid evolution of LLMs, staying informed about new developments and maintaining a proactive approach to model management is essential for leveraging the full potential of Generative AI.

Deploying LLMs is not just about selecting the best model; it's about creating a comprehensive strategy that integrates the model into a robust, scalable, and secure system. By doing so, organizations can unlock the transformative power of Generative AI, driving innovation and efficiency across a wide range of industries and applications.

Chapter 5

Design & Architecture Principles for Generative AI Solutions

Generative AI (GenAI) solutions have rapidly evolved, and as they become more complex and widely adopted, it is essential to establish sound design and architecture principles. These principles ensure that GenAI solutions are not only efficient and scalable but also ethical, secure, and aligned with business objectives. This chapter covers key principles to consider in the design and architecture of a robust, responsible, and future-ready GenAI system.

1. Scalability and Modularity

Explanation

Scalability and modularity are critical for GenAI solutions, as these models can require massive computational resources, especially for training and inference tasks. As the solution grows in data volume, user requests, or model complexity, it must handle increased workloads without compromising performance. Modularity allows components of the GenAI solution to be developed, maintained, and scaled independently.

Key Considerations

- Vertical and Horizontal Scaling: Implement vertical scaling (upgrading hardware) for simpler systems and horizontal scaling (adding more nodes) for distributed systems. Use cloud resources that allow auto-scaling based on demand.
- Microservices Architecture: Decompose the GenAI solution into smaller, loosely coupled services. For example, have separate services for data ingestion, model training, model inference, and post-processing.

- Serverless and Containerization: Use serverless architectures (e.g., AWS Lambda, Azure Functions) or containerization (e.g., Docker, Kubernetes) to simplify scaling and manage resource allocation efficiently.

Key Use Cases

- Real-Time Language Translation: Real-time applications require dynamic scaling to manage spikes in demand.
- Image Generation on Demand: Modular design enables handling of high loads for user-generated requests.

Cloud Tools

- AWS: Use Amazon Elastic Kubernetes Service (EKS) and AWS Lambda for serverless functions. Amazon SageMaker provides managed infrastructure for scalable model training and deployment.
- Google Cloud: Google Kubernetes Engine (GKE) supports containerized, scalable architectures. Vertex AI offers a modular platform for building, deploying, and scaling ML models.
- Azure: Azure Kubernetes Service (AKS) and Azure Functions enable scalable, serverless architectures. Azure Machine Learning provides an environment for building and managing scalable GenAI models.

Best Practices

- Design stateless services where possible, as they are easier to scale.
- Use load balancers to distribute workloads and manage traffic effectively.
- Ensure each module or microservice has a well-defined API, facilitating independent development and updates.

2. Data Privacy and Security

Explanation

Data privacy and security are fundamental when designing GenAI solutions, especially when dealing with sensitive data. Strong privacy and security measures protect the solution from unauthorized access, data leaks, and breaches, ensuring compliance with regulations like GDPR and CCPA.

Key Considerations

- Data Encryption: Encrypt all data, both at rest and in transit, using industry-standard encryption methods to prevent unauthorized access.
- Access Control and Authentication: Implement strict access control policies. Use multi-factor authentication (MFA) and role-based access control (RBAC) to restrict access to sensitive components.
- Data Anonymization and Masking: For sensitive datasets, use anonymization techniques that mask personally identifiable information (PII) while preserving data utility.

Key Use Cases

- Healthcare Data Analysis: Ensures privacy of patient data during AI-driven medical analysis.
- Financial Fraud Detection: Protects sensitive financial data in fraud detection applications.

Cloud Tools

- AWS: AWS Key Management Service (KMS) for encryption and AWS Identity and Access Management (IAM) for access control. Amazon Macie for discovering and protecting sensitive data.
- Google Cloud: Cloud Data Loss Prevention (DLP) for detecting and masking sensitive data, Cloud IAM for fine-grained access control, and Cloud KMS for encryption.
- Azure: Azure Key Vault for managing encryption keys, Azure Policy for enforcing security standards, and Azure AD for identity and access management.

Best Practices

- Regularly audit data access logs to detect and address unauthorized activities.
- Use differential privacy techniques to provide insights without compromising individual privacy.
- Implement security best practices across the system, including securing APIs, setting up firewalls, and conducting vulnerability assessments.

3. Ethics and Bias Mitigation

Explanation

Generative AI systems can perpetuate or even amplify biases present in training data, leading to unfair or unethical outcomes. Ethics and bias mitigation principles are critical to ensuring that the AI model is fair, transparent, and responsibly used.

Key Considerations

- Bias Detection and Mitigation: Perform regular checks to detect biases in the training data or model outputs, especially in applications like hiring, lending, or law enforcement.
- Explainability and Transparency: Use Explainable AI (XAI) techniques to make the model's decision-making process interpretable, particularly for stakeholders and users.
- Fairness Audits: Regularly conduct fairness audits to assess the impact of the model across different demographic groups and ensure equitable treatment.

Key Use Cases

- Hiring and Recruitment: Reducing bias in resume screening and candidate assessments.
- Loan Approval: Preventing unfair treatment in financial decision-making.

Cloud Tools

- AWS: Amazon SageMaker Clarify for detecting bias in data and models, and providing transparency into model predictions.
- Google Cloud: What-If Tool in AI Platform for analyzing model behavior and bias. Explainable AI in Vertex AI helps ensure transparency.
- Azure: Fairlearn in Azure Machine Learning for fairness assessments, and InterpretML for explainability of model predictions.

Best Practices

- Include diverse datasets that represent different population segments to reduce the risk of bias.
- Implement clear documentation that explains how the model was trained, its limitations, and its intended use.
- Establish a feedback loop for users to report issues, which can be analyzed and used to improve the model.

4. Adaptability and Flexibility

Explanation

GenAI solutions should be adaptable to various use cases and flexible enough to incorporate new advancements in AI technologies. This flexibility enables the solution to stay relevant and cater to evolving business needs.

Key Considerations

- Configurable Models: Allow parameters and configurations to be easily adjusted to meet different requirements without extensive retraining.
- Model Agnosticism: Design the architecture to support multiple model types (e.g., GPT, BERT, DALL-E) and allow switching or upgrading to newer models as they become available.
- Continuous Learning: Establish a framework for continuous learning, where the model can be retrained or fine-tuned as new data becomes available.

Key Use Cases

- Customer Service Chatbots: Adapts to various industries and evolving user needs.
- Personalized Content Generation: Adjusts content based on user profiles or preferences.

Cloud Tools

- AWS: Amazon SageMaker provides support for a variety of ML frameworks, making it adaptable for different models. AWS Lambda allows for flexible, event-driven processing.
- Google Cloud: Vertex AI enables easy switching between model types and frameworks. Cloud Functions offer event-driven flexibility.

- Azure: Azure Machine Learning supports multiple ML frameworks. Azure Logic Apps facilitate flexible, event-driven workflows.

Best Practices

- Design the solution so that different model versions can coexist, enabling A/B testing and comparison.
- Use an MLOps pipeline to automate model retraining, versioning, and deployment, ensuring quick adaptability.
- Maintain modularity in data processing, model inference, and post-processing, allowing easy customization for different tasks.

5. Efficiency and Optimization

Explanation

GenAI models are resource-intensive, requiring significant computational power for both training and inference. Efficiency and optimization principles focus on reducing resource consumption and operational costs while maximizing model performance.

Key Considerations

- Resource Optimization: Use model pruning, quantization, and other techniques to reduce the model size and improve performance without sacrificing accuracy.
- Cost-Efficient Infrastructure: Leverage cloud providers' cost-saving features, such as spot instances, to lower operational costs.
- Inference Optimization: Implement caching mechanisms for frequently used results and use lower-precision formats (e.g., FP16) for inference where possible.

Key Use Cases

- Real-Time Image Generation: Optimization is necessary to generate high-quality images quickly and cost-effectively.
- Natural Language Processing (NLP): Reducing inference times for real-time language models.

Cloud Tools

- AWS: Amazon EC2 Spot Instances and SageMaker Neo for cost-effective, optimized machine learning inference.
- Google Cloud: TPU Pods for efficient training of large models, and Vertex AI Optimizer for fine-tuning model performance.
- Azure: Azure Spot VMs for cost savings on compute resources, and Azure Machine Learning Managed Endpoints for efficient model deployment.

Best Practices

- Profile the model's performance regularly to identify and address bottlenecks.
- Use batch processing and asynchronous workflows to handle large volumes of requests efficiently.
- Regularly update models with the latest research advances in efficient architecture, like transformer distillation or sparse transformers.

6. Observability and Monitoring

Explanation

Observability and monitoring ensure that the GenAI solution is functioning as expected. They provide insights into system health, performance, and model accuracy, enabling proactive issue detection and troubleshooting.

Key Considerations

- Performance Metrics: Track metrics such as latency, response time, accuracy, and error rates to ensure optimal performance.
- Anomaly Detection: Implement anomaly detection to identify unusual patterns or model drift that may impact performance or accuracy.
- End-to-End Monitoring: Use logging, tracing, and monitoring tools to capture metrics across all stages of the pipeline, from data ingestion to model inference.

Key Use Cases

- Fraud Detection Systems: Monitors model performance to identify drifts that could impact accuracy.
- Recommendation Engines: Observability ensures the system responds appropriately to changes in user behavior.

Cloud Tools

- AWS: Amazon CloudWatch for monitoring and logging, SageMaker Model Monitor for detecting data drift and model quality issues.
- Google Cloud: Cloud Monitoring and Cloud Logging provide observability for applications, while Vertex AI Model Monitoring tracks ML model performance.
- Azure: Azure Monitor and Application Insights for end-to-end monitoring, Azure ML Model Data Collector for tracking model performance.

Best Practices

- Set up alerting mechanisms for critical metrics, allowing rapid response to potential issues.
- Regularly review logs to identify patterns that may indicate security risks or performance degradation.
- Use AIOps (Artificial Intelligence for IT Operations) to automate the detection of issues and recommend actions for troubleshooting.

7. User-Centric Design

Explanation

User-centric design ensures that the GenAI solution is intuitive, useful, and meets end-user needs. This principle emphasizes designing with the user in mind, prioritizing accessibility, ease of use, and providing relevant outputs.

Key Considerations

- Human-Centered Interaction: Ensure that the solution's outputs are understandable and actionable for users. Avoid jargon and make results easy to interpret.
- Feedback Mechanisms: Include mechanisms for users to provide feedback on model outputs, allowing for iterative improvement.
- Personalization: Implement features that personalize responses based on individual preferences, where appropriate.

Key Use Cases

- Virtual Assistants: Requires user-friendly design to enhance usability and accessibility.
- Customer Support Bots: Provides clear responses and gathers user feedback to improve interaction quality.

Cloud Tools

- AWS: Amazon Lex for building conversational interfaces and Amazon Personalize for personalized user experiences.
- Google Cloud: Dialogflow for conversational agents, Recommendations AI for personalized recommendations.
- Azure: Azure Bot Services for building and deploying chatbots, Azure Personalizer for real-time personalization.

Best Practices

- Conduct user testing to understand how users interact with the solution and identify areas for improvement.
- Regularly collect and analyze user feedback to refine the model's behavior.
- Design user interfaces that clearly explain the AI's role, limitations, and actions it can perform.

8. Compliance and Governance

Explanation

Compliance and governance ensure that the GenAI solution meets regulatory requirements and follows organizational policies, especially concerning data privacy, security, and ethical considerations.

Key Considerations

- Adherence to Regulations: Ensure compliance with industry standards and legal requirements, such as GDPR, CCPA, HIPAA, and sector-specific guidelines.
- Audit Trails: Maintain detailed logs of data usage, model decisions, and updates to support compliance and traceability.
- Model Governance: Implement a governance framework to track model versions, monitor performance, and oversee responsible use.

Key Use Cases

- Healthcare Diagnostics: Ensures compliance with HIPAA and other regulations.
- Financial Services: Adheres to regulatory standards in banking and financial services.

Cloud Tools

- AWS: AWS Config for monitoring compliance, SageMaker Model Governance for model tracking, and AWS Audit Manager.
- Google Cloud: Policy Intelligence for compliance, Cloud DLP for data protection, and AI Platform Governance.
- Azure: Azure Policy for enforcing compliance rules, Azure Purview for data governance, and Azure Blueprints for governance templates.

Best Practices

- Conduct regular compliance audits to ensure adherence to changing regulations.
- Document the end-to-end lifecycle of the model, including training, deployment, and retirement.
- Establish clear policies for data usage, model updates, and handling of model outputs.

9. High Availability and Fault Tolerance

Explanation

High availability and fault tolerance are essential for GenAI solutions, especially those that support mission-critical functions. These principles focus on ensuring the solution remains available and resilient, even in the face of infrastructure issues or unexpected spikes in demand.

Key Considerations

- Redundancy: Design redundancy into infrastructure components to reduce single points of failure.
- Failover Mechanisms: Implement failover mechanisms to ensure continuity in case of a component failure.
- Disaster Recovery: Develop a disaster recovery plan with regular backups and clear protocols for restoring service.

Key Use Cases

- E-commerce Chatbots: High availability ensures continuous customer support.
- Medical Diagnostics: Fault tolerance ensures critical services are available in emergencies.

Cloud Tools

- AWS: Amazon RDS Multi-AZ for database failover, AWS Elastic Load Balancing (ELB) for distributing traffic, and AWS Backup for disaster recovery.
- Google Cloud: Global Load Balancer for high availability, Cloud SQL High Availability, and Cloud Backup.
- Azure: Azure Site Recovery for disaster recovery, Load Balancer, and Azure SQL with Failover Groups.

Best Practices

- Test failover and recovery processes regularly to ensure readiness.
- Use load balancing to distribute traffic and prevent overloading specific components.
- Implement multi-zone or multi-region deployments for critical applications to enhance resilience.

10. Lifecycle Management and MLOps

Explanation

Lifecycle management and MLOps (Machine Learning Operations) are vital for maintaining, updating, and scaling GenAI models. This principle ensures that the solution can adapt over time and remain effective in changing environments.

Key Considerations

- Model Versioning: Track different versions of the model to manage updates and rollbacks effectively.
- Continuous Integration/Continuous Deployment (CI/CD): Use CI/CD pipelines for rapid, reliable deployment of model updates.
- Automated Retraining: Set up pipelines for automated retraining, so the model stays current with new data and performs consistently.

Key Use Cases

- Automated Document Generation: Requires periodic updates to incorporate new document types and formats.
- Dynamic Pricing Models: Regular updates based on market trends and customer data.

Cloud Tools

- AWS: Amazon SageMaker Pipelines for MLOps, AWS CodePipeline for CI/CD, and SageMaker Model Registry for versioning.
- Google Cloud: Vertex AI Pipelines for MLOps workflows, Cloud Build for CI/CD, and Vertex Model Registry.

- Azure: Azure Machine Learning Pipelines for model lifecycle management, Azure DevOps for CI/CD, and Azure ML Model Registry.

Best Practices

- Regularly evaluate model performance and retrain as needed to address model drift.
- Implement model monitoring to detect when the model needs updating.
- Maintain a record of all changes made to the model for accountability and transparency.

11. Interoperability and Integration

Explanation

Interoperability and integration ensure that the GenAI solution can communicate effectively with other systems and applications within an organization's ecosystem.

Key Considerations

- Standardized APIs: Design APIs to facilitate seamless data exchange between the GenAI solution and other systems.
- Deployment Flexibility: Ensure the solution can be deployed on-premises, in the cloud, or in hybrid environments based on organizational requirements.
- Compatibility with Existing Systems: Design the GenAI solution to work with existing data sources, analytics platforms, and operational systems.

Key Use Cases

- Multi-Platform Recommendation Systems: Integrates across web and mobile platforms to provide a seamless user experience.
- Data-Driven Insights: Integrates AI insights with business intelligence platforms for comprehensive reporting.

Cloud Tools

- AWS: Amazon API Gateway for API management, AWS Glue for data integration, and Amazon EventBridge for event-driven integrations.
- Google Cloud: Apigee API Management for APIs, Cloud Data Fusion for data integration, and Pub/Sub for event-based architecture.
- Azure: Azure API Management, Azure Data Factory for data integration, and Event Grid for event-driven processing.

Best Practices

- Use RESTful or GraphQL APIs for standardized, flexible integrations.
- Regularly update integration points to ensure compatibility with evolving systems.
- Implement data transformation pipelines to align data formats across systems.

Conclusion

By adhering to these design and architecture principles, organizations can create GenAI solutions that are scalable, efficient, secure, and aligned with ethical and business objectives. These principles serve as guardrails, ensuring that GenAI applications are not only powerful but also responsible and sustainable in the long run.

Chapter 6

GenAI Reference Architectures

This reference architecture provides a framework for building robust GenAI systems. Depending on specific use cases or requirements, some components might be more emphasized or additional elements might be added This diagram represents a high-level GenAI Reference Architecture. Let us discuss the key components:

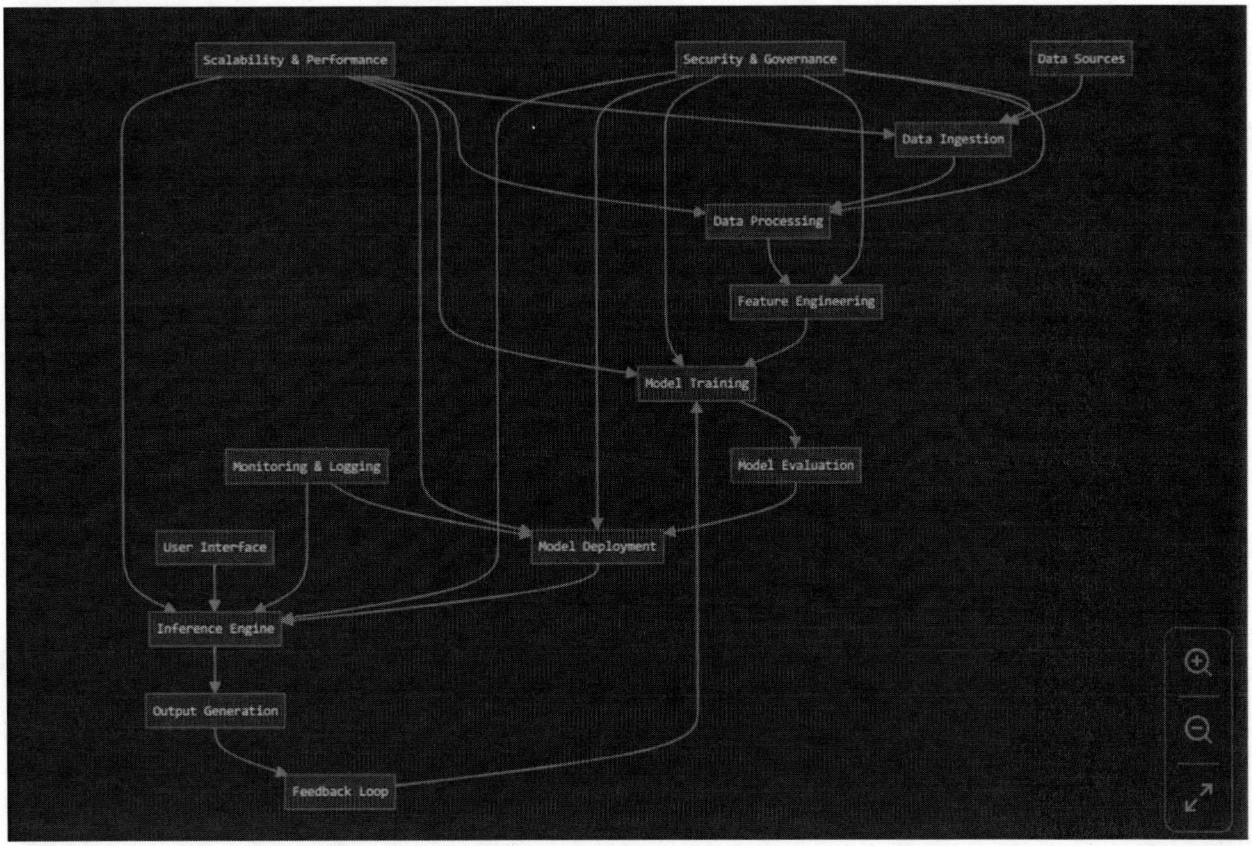

1. Data Sources: Various inputs like text, images, audio, or structured data.
2. Data Ingestion: Collecting and importing data from various sources.
3. Data Processing: Cleaning, normalizing, and preparing the data for model training.
4. Feature Engineering: Extracting and creating relevant features from the processed data.
5. Model Training: Using the prepared data to train the generative AI model.
6. Model Evaluation: Assessing the model's performance and fine-tuning as needed.
7. Model Deployment: Putting the trained model into production.
8. Inference Engine: The component that uses the deployed model to generate outputs.
9. User Interface: Where users interact with the system to input prompts or requests.
10. Output Generation: The final step where the AI generates the requested content.
11. Monitoring & Logging: Continuous tracking of system performance and user interactions.
12. Feedback Loop: Mechanism to incorporate user feedback and improve the model over time.
13. Security & Governance: Ensuring data privacy, model fairness, and ethical use.
14. Scalability & Performance: Infrastructure to handle increasing loads and maintain speed.

1. GenAI Reference Architecture AWS

Creating a robust GenAI (Generative AI) reference architecture on AWS involves leveraging a combination of AWS services, best practices, and tools to build, train, and deploy machine learning models effectively. The architecture can be structured to support various AI use cases such as text generation, image creation, chatbots, and more. Below is a high-level design for an AWS GenAI Reference Architecture:

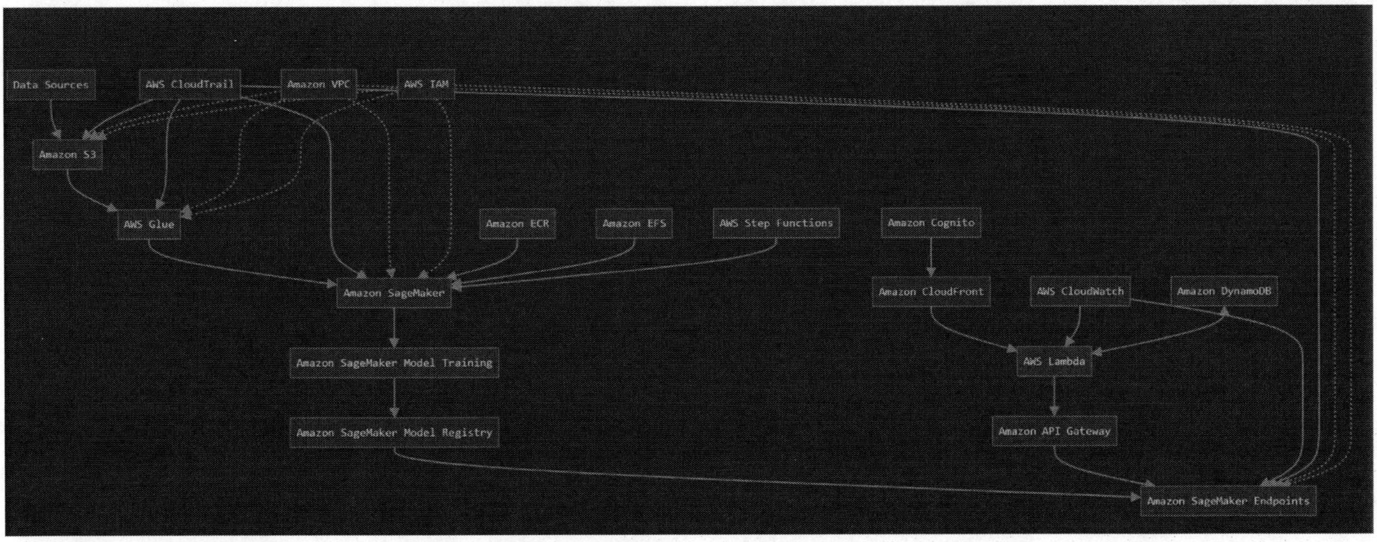

AWS GenAI Reference Architecture

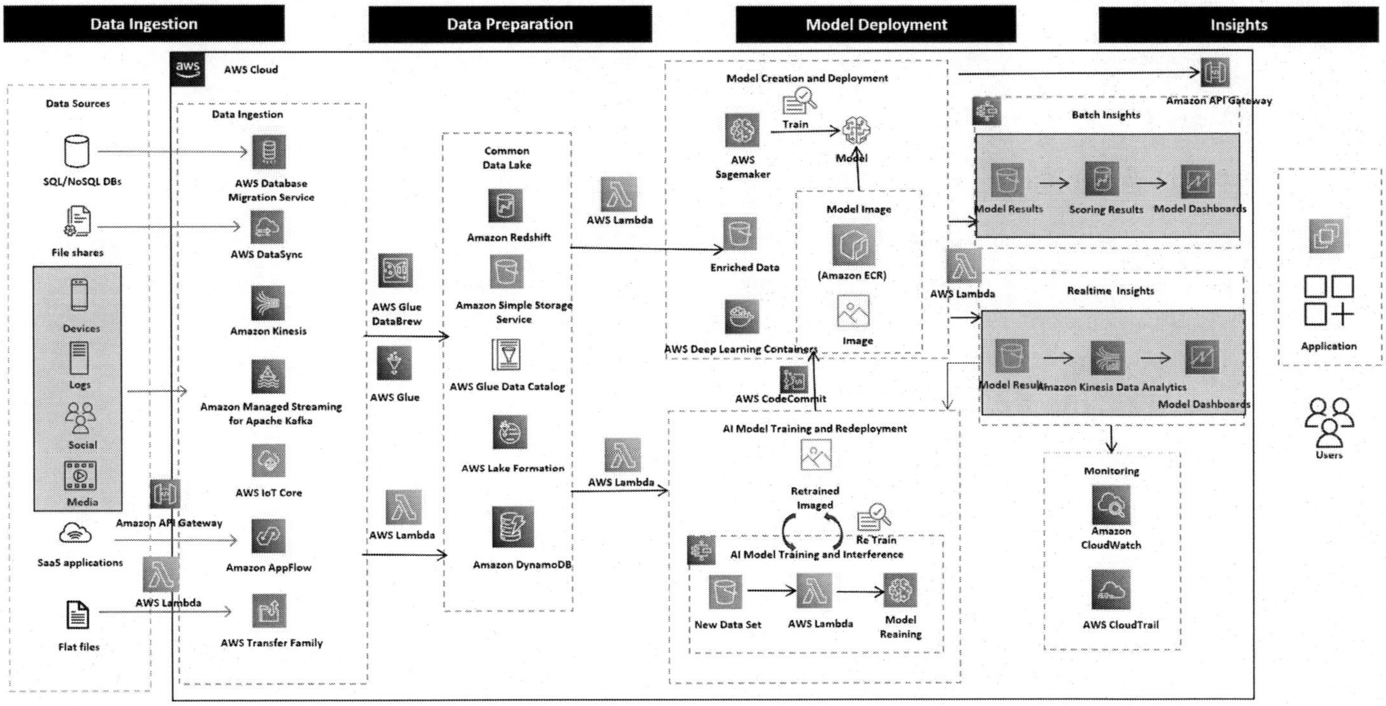

1.1 Data Ingestion and Storage Layer

This layer handles the ingestion and storage of data needed for training and serving the models.

- **Amazon S3 (Simple Storage Service)**: Centralized data lake to store raw, processed, and feature-engineered data.
- **AWS Glue**: For data cataloging and ETL (Extract, Transform, Load) processes.
- **Amazon Kinesis / AWS Lambda**: Stream processing for real-time data ingestion.
- **Amazon RDS / DynamoDB**: Structured data storage for transactional or metadata needs.

1.2. Data Processing and Feature Engineering Layer

Processing and transforming raw data into a suitable format for model training.

- **AWS Glue / AWS Lambda**: To perform batch and real-time data transformations.
- **Amazon EMR / AWS Batch**: For large-scale distributed data processing.
- **Amazon SageMaker Data Wrangler**: For feature engineering and data preparation.
- **Amazon Athena**: Serverless querying to explore and analyze data.

1.3. Model Training and Experimentation Layer

This layer focuses on building, training, and experimenting with generative models.

- **Amazon SageMaker Studio**: Central IDE for building, training, and deploying models.
- **Amazon SageMaker Training Jobs**: For running distributed training on large datasets.
- **Amazon SageMaker Experiments**: To track and manage various training runs and hyperparameters.
- **Amazon EC2 Instances (P3/P4 Instances)**: For custom training on GPU/TPU instances.
- **AWS ParallelCluster**: For scalable training using HPC (High-Performance Computing) clusters.
- **Amazon FSx for Lustre**: High-performance storage for fast data access during training.

1.4. Model Deployment and Serving Layer

Deploy trained models for inference and integrate with downstream applications.

- **Amazon SageMaker Endpoints**: For real-time model hosting and serving.
- **Amazon ECS / EKS (Elastic Container Service / Kubernetes Service)**: For containerized deployment of custom GenAI models.
- **AWS Lambda**: For serverless, low-latency inference tasks.
- **Amazon API Gateway**: To create secure RESTful APIs for model endpoints.
- **AWS App Runner**: To quickly deploy and scale GenAI models without managing servers.

1.5. MLOps and Model Monitoring Layer

Automate CI/CD pipelines, monitor model performance, and retrain models as needed.

- **Amazon SageMaker Pipelines**: For creating robust MLOps pipelines, including data preprocessing, training, and deployment.
- **AWS CodePipeline / CodeBuild / CodeCommit**: For CI/CD orchestration and code management.
- **Amazon SageMaker Model Monitor**: To track data and model drift, ensuring models perform well on live data.
- **Amazon CloudWatch / AWS X-Ray**: For logging, monitoring, and tracing model inference.

- **Amazon SageMaker Clarify**: For detecting bias in training data and models.

1.6. Security, Compliance, and Governance Layer

Ensuring data protection, compliance, and governance throughout the AI lifecycle.

- **AWS IAM (Identity and Access Management)**: Fine-grained access control for resources.
- **AWS KMS (Key Management Service)**: To encrypt data at rest and in transit.
- **AWS Secrets Manager / Systems Manager Parameter Store**: To securely manage and access secrets like API keys and credentials.
- **AWS CloudTrail**: For auditing and compliance.

1.7. GenAI Application and User Interaction Layer

Enable end-users and downstream applications to interact with the Generative AI model outputs.

- **Amazon Lex**: For building conversational AI interfaces (chatbots).
- **Amazon Polly**: For converting text outputs to lifelike speech.
- **Amazon Rekognition**: To integrate computer vision capabilities.
- **Amazon Transcribe / Translate**: For multi-language support.
- **AWS Amplify / AWS AppSync**: To create and manage front-end user applications.

1.8. Cost Optimization and Scaling Layer

Design for scalability and optimize costs based on usage patterns.

- **Amazon SageMaker Serverless Inference**: Cost-effective for sporadic workloads.
- **AWS Auto Scaling**: For dynamically adjusting resources based on traffic.
- **AWS Compute Optimizer**: For recommendations on EC2 and SageMaker instances.
- **AWS Trusted Advisor**: To ensure adherence to best practices

2 GenAI Reference Architecture Google

A high-level Google Cloud GenAI (Generative AI) reference architecture. This architecture is designed for implementing a generative AI solution leveraging Google Cloud's services. It covers the various components required for building, training, and deploying generative models.

This architecture provides a comprehensive guide for building, deploying, and maintaining GenAI models on Google Cloud. Depending on the use case, additional components or services (e.g., Anthos for hybrid deployments, or Google Cloud's AI Building Blocks for vision and translation) can be added.

Here's a brief overview of the flow:

1. Data is ingested into Cloud Storage and processed using Dataflow.
2. Processed data is stored in BigQuery for analysis and feature engineering.
3. Vertex AI is used for model training, management, and deployment, potentially leveraging pre-trained models via PaLM API.
4. Cloud Run and Cloud Functions handle user requests, which are served by Vertex AI Prediction.
5. Cloud Load Balancing ensures efficient distribution of incoming traffic.
6. Firebase Authentication manages user authentication.
7. The entire system is monitored using Cloud Monitoring and Cloud Audit Logs, with security ensured by VPC and IAM.

This architecture provides a scalable, secure, and flexible framework for building GenAI solutions on Google Cloud. It can be customized based on specific requirements, such as data volume, model complexity, or specific security needs.

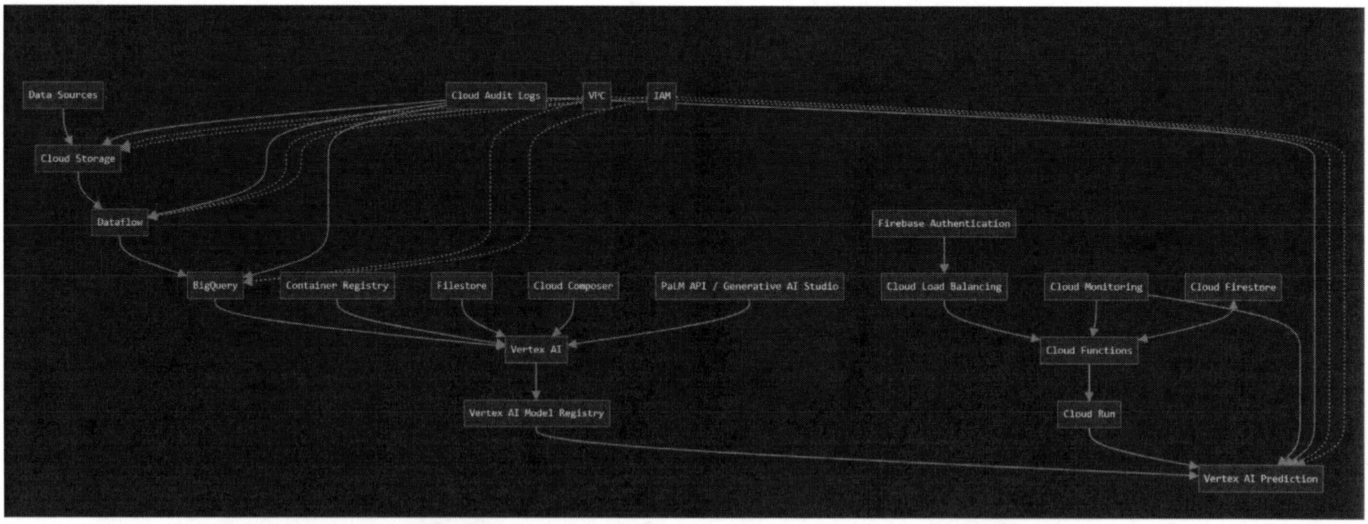

1. **Detailed Component Breakdown and Best Practices**
   - AI and Machine Learning Services
     - **Vertex AI:** Unified ML platform for the entire machine learning lifecycle.
     - **Key Features**:
       1. AutoML for automated model training
       2. Custom model support
       3. Model deployment and monitoring

    4. Feature store for managing ML features
  - **Best Practices:**
    1. Use Vertex AI Workbench for collaborative development
    2. Leverage Vertex AI Pipelines for reproducible ML workflows
    3. Implement Vertex AI Model Monitoring for production models
    4. Utilize Vertex AI Feature Store for consistent feature management
- **Cloud TPU:** Specialized hardware accelerators for ML workloads.
  - **Key Features:**
    1. High-performance matrix multiplication
    2. Scalable from single devices to POD slices
    3. Integration with TensorFlow and PyTorch
  - **Best Practices:**
    1. Use TPU VMs for direct access to TPU devices
    2. Optimize models for TPU architecture
    3. Leverage TPU Pod slices for large-scale training

- **BigQuery ML:** ML model training and inference directly in BigQuery.
  - **Key Features:**
    1. SQL-based model creation and training
    2. Support for various model types (linear regression, logistic regression, etc.)
    3. Integration with BigQuery's data processing capabilities
  - **Best Practices:**
    1. Use for quick prototyping and simple models
    2. Leverage BigQuery ML for feature engineering at scale
    3. Implement model iterations using BigQuery ML's versioning
- **Data Storage and Processing**
  - **Cloud Storage:** Scalable object storage for any amount of data.
  - **Key Features:**
    1. Multiple storage classes (Standard, Nearline, Coldline, Archive)
    2. Strong consistency and geo-redundancy
    3. Integration with other GCP services
  - **Best Practices:**
    1. Implement lifecycle policies for cost-effective data management
    2. Use signed URLs for secure, temporary object access
    3. Leverage Cloud Storage Transfer Service for large-scale data ingestion
- **b. BigQuery:** Serverless, scalable data warehouse.
  - **Key Features:**
    1. Petabyte-scale querying
    2. Separation of storage and compute
    3. ML integration with BigQuery ML
  - **Best Practices:**

4. Implement proper table partitioning and clustering
5. Use authorized views for fine-grained access control
6. Leverage BigQuery BI Engine for fast, interactive dashboards

- **c. Dataflow:** Unified stream and batch data processing.
  - **Key Features:**
    1. Auto-scaling and dynamic work rebalancing
    2. Support for complex data processing patterns
    3. Integration with other GCP services
  - **Best Practices:**
    1. Use Dataflow templates for common processing patterns
    2. Implement windowing strategies for streaming data
    3. Leverage Dataflow Prime for improved performance and cost-efficiency
- **Compute and Orchestration**
  - **Cloud Functions: Event-driven, serverless compute platform.**
  - **Key Features:**
    1. Automatic scaling and management
    2. Pay-per-use pricing
    3. Support for multiple programming languages
  - **Best Practices:**
    1. Use for lightweight, event-driven tasks
    2. Implement proper error handling and retries
    3. Leverage Cloud Functions for ETL tasks and data transformations
- **b. Cloud Run:** Fully managed platform for containerized applications.
  - **Key Features:**
    1. Auto-scaling to zero
    2. Support for any programming language
    3. Integration with Cloud Build and Container Registry
  - **Best Practices:**
    1. Optimize container images for faster startup
    2. Implement health checks for improved reliability
    3. Use Cloud Run for serving ML models and APIs

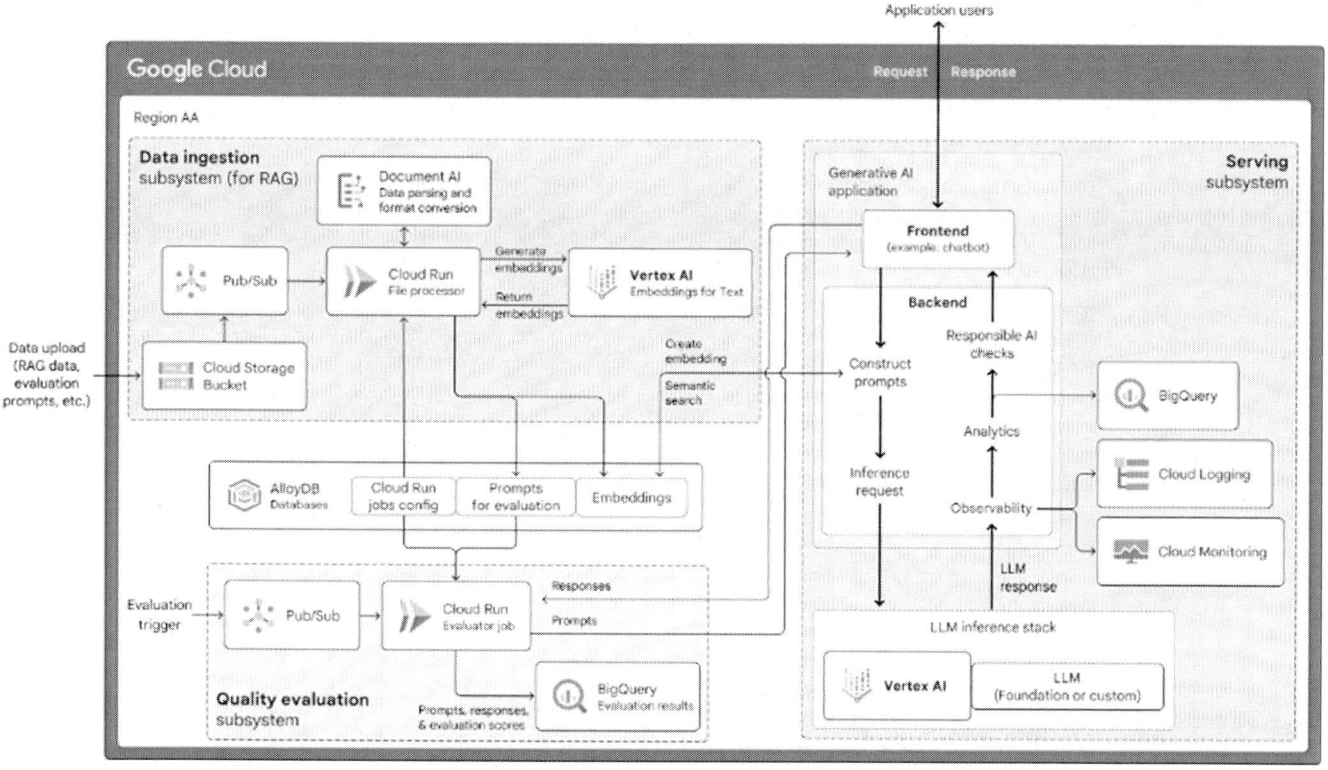

***The above picture has been taken from https://cloud.google.com/static/architecture/images/rag-capable-gen-ai-app-using-vertex-ai-architecture-detail.svg

- **c. Kubernetes Engine (GKE):** Managed Kubernetes service.
  - **Key Features:**
    1. Auto-upgrade and auto-repair
    2. Integration with GCP services
    3. Support for both stateless and stateful applications
  - **Best Practices:**
    1. Use GKE Autopilot for fully managed Kubernetes
    2. Implement proper resource requests and limits
    3. Leverage GKE's built-in logging and monitoring
- **Networking and Security**
  - **Cloud VPC: Software-defined networking for GCP resources.**
  - **Key Features:**
    1. Global network with regional subnetworks
    2. VPC peering and Shared VPC
    3. Firewall rules and network tags
  - **Best Practices:**
    1. Implement proper network segmentation
    2. Use VPC Service Controls for additional data security
    3. Leverage Cloud NAT for outbound internet access

- **b. Cloud Armor:** Web application firewall and DDoS protection.
  - **Key Features:**
    1. Pre-configured rules for OWASP Top 10 risks
    2. Custom rule creation
    3. Integration with Cloud CDN
  - **Best Practices:**
    1. Implement both pre-configured and custom rules
    2. Use Cloud Armor's adaptive protection to detect and mitigate sophisticated attacks
    3. Regularly review and update security policies
- **c. Identity and Access Management (IAM):** Fine-grained access control for GCP resources.
  - **Key Features:**
    1. Predefined and custom IAM roles
    2. Service accounts for machine-to-machine interactions
    3. Integration with Cloud Identity for user management
  - **Best Practices:**
    1. Implement the principle of least privilege
    2. Use service accounts with appropriate permissions for applications
    3. Regularly audit and review IAM policies
- Monitoring and Operations
  - **Cloud Monitoring:** Comprehensive monitoring solution for GCP and AWS resources.
  - **Key Features:**
    1. Custom dashboards and alerting
    2. Uptime checks and synthetic monitoring
    3. Integration with third-party monitoring tools
  - **Best Practices:**
    4. Create custom dashboards for key metrics
    5. Implement proactive alerting for critical issues
    6. Use workspaces to organize resources and manage access
- **b. Cloud Logging:** Fully managed, real-time log management and analysis.
  - **Key Features:**
    1. Centralized log storage and analysis
    2. Integration with BigQuery and Pub/Sub
    3. Log-based metrics creation
  - **Best Practices:**
    1. Implement structured logging for easier analysis
    2. Use log sinks to export logs to other GCP services
    3. Leverage log-based metrics for custom monitoring
- **c. Error Reporting:** Real-time exception monitoring and alerting.
  - **Key Features:**
    1. Automatic error grouping
    2. Integration with popular error tracking libraries
    3. Notifications via email, mobile push, and Pub/Sub

- **Best Practices:**
  1. Implement proper error handling in applications
  2. Set up alerts for critical errors
  3. Use Error Reporting in conjunction with Cloud Debugger for faster issue resolution

3 Gen AI Reference Architecture Azure

Microsoft Azure offers a comprehensive suite of tools and services to build, train, deploy, and monitor Generative AI models. Whether it's natural language processing (NLP), image generation, or multimodal applications, Azure's cloud-native ecosystem provides an integrated environment for building large-scale AI solutions. This chapter presents a detailed reference architecture for implementing Generative AI on Azure, covering key services, components, and best practices for deploying end-to-end AI solutions.

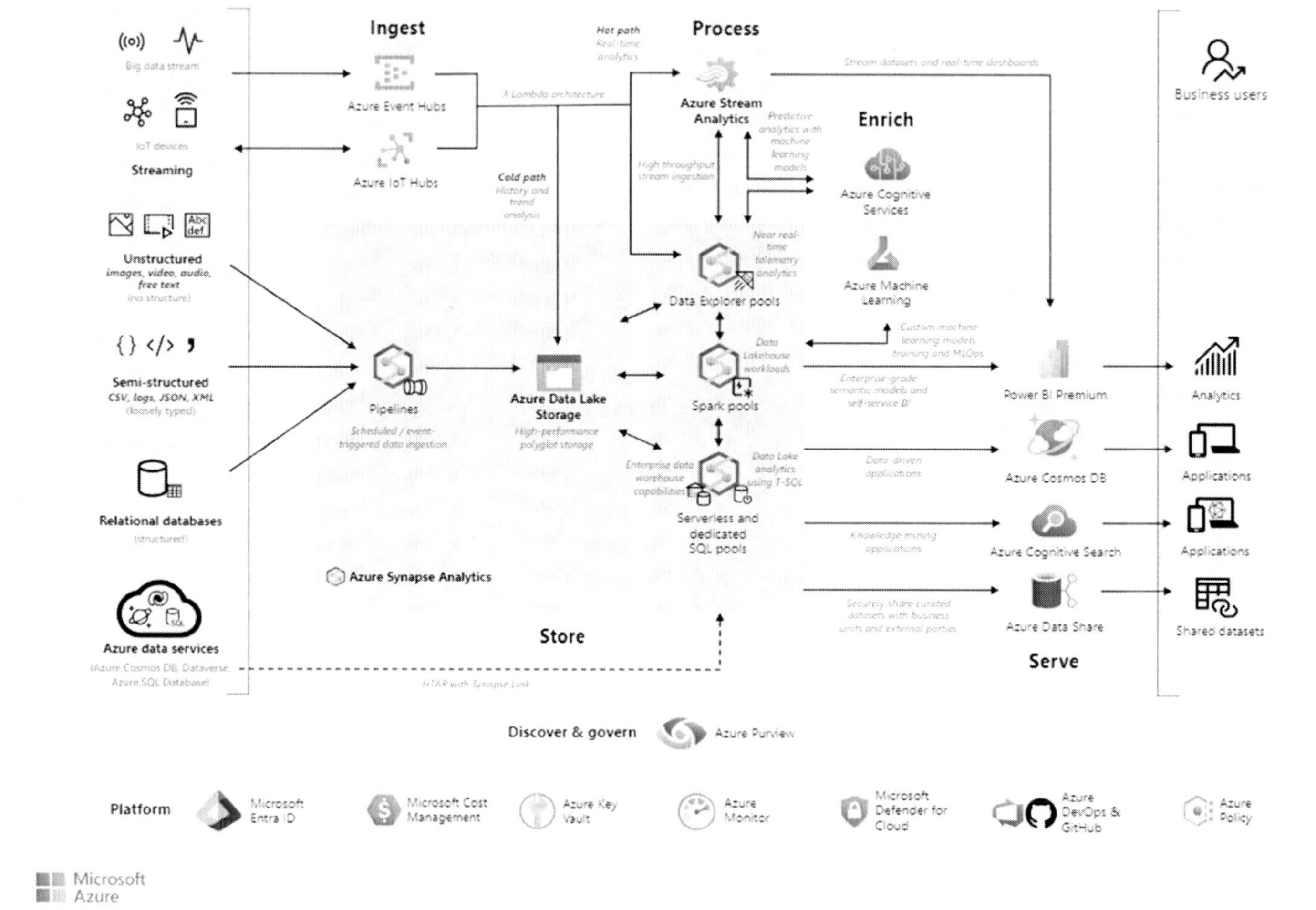

*** - The above image has been taken from https://learn.microsoft.com/en-us/azure/architecture/example-scenario/dataplate2e/media/azure-analytics-end-to-end.svg

The following section elaborate on each part of the architecture, including relevant Azure services and best practices for implementation:

1. Client Applications : These could be web, mobile, or desktop applications that interact with the AI services
   - Services: Web Apps, Mobile Apps, Desktop Apps
   - Best Practices:
     - Use responsive design for web applications
     - Implement proper error handling and user feedback
     - Utilize Azure AD for authentication and authorization
2. Azure API Management: Acts as a gateway for managing and securing APIs.
   - Services: Azure API Management

- Best Practices:
  - Implement rate limiting and quotas
  - Use OAuth 2.0 or OpenID Connect for API security
  - Set up custom domains and SSL certificates
  - Implement API versioning

3. Azure Functions: Serverless compute service for running your AI logic and orchestrating different services.
   - Services: Azure Functions
   - Best Practices:
     - Use durable functions for complex orchestrations
     - Implement proper error handling and logging
     - Use managed identities for secure access to other Azure resources
     - Optimize function performance with async programming
4. Azure OpenAI Service: Provides access to OpenAI's powerful language models like GPT-4.
   - Services: Azure OpenAI Service
   - Best Practices:
     - Use the latest model versions when appropriate
     - Implement content filtering to ensure safe outputs
     - Use fine-tuning for domain-specific tasks
     - Implement prompt engineering techniques for better results

5. Azure Cognitive Search: Enables intelligent search capabilities over your data.
   - Services: Azure Cognitive Search
   - Best Practices:
     - Use semantic search for more relevant results
     - Implement faceted navigation for better user experience
     - Use custom skillsets to enrich your data
     - Optimize indexing for large datasets
6. Azure Cognitive Services : Offers pre-built AI models for various tasks like vision, speech, and language understanding.
   - Services: Computer Vision, Speech Services, Language Understanding
   - Best Practices:
     - Use containerized versions for on-premises or edge deployments
     - Implement retry logic for API calls
     - Use custom models when pre-built models don't meet requirements
7. Azure Machine Learning: For custom model training and deployment.
   - Services: Azure Machine Learning
   - Best Practices:
     - Use MLOps practices for model lifecycle management
     - Implement automated ML for model selection and hyperparameter tuning
     - Use distributed training for large models

    - Implement model monitoring and retraining pipelines

8. Azure Storage : Stores data used by the AI services and models.
    - Services: Blob Storage, Azure Data Lake Storage
    - Best Practices:
        - Use appropriate storage tiers for cost optimization
        - Implement proper data lifecycle management
        - Use Azure Private Link for secure access
        - Enable soft delete and versioning for data protection
9. Azure Monitor : Provides monitoring and logging capabilities across all services.
    - Services: Azure Monitor, Application Insights
    - Best Practices:
        - Set up custom dashboards for key metrics
        - Implement alerts for critical issues
        - Use Log Analytics for advanced querying and analysis
        - Enable continuous export of telemetry data
10. Azure Key Vault: Securely stores and manages sensitive information like API keys and secrets
    - Services: Azure Key Vault
    - Best Practices:
        - Use managed identities for secure access
        - Implement proper RBAC for key vault access
        - Enable soft-delete and purge protection
        - Regularly rotate secrets and keys
11. Azure Active Directory (New)
    - Services: Azure Active Directory
    - Best Practices:
        - Implement Multi-Factor Authentication
        - Use Conditional Access policies
        - Enable Identity Protection features
        - Implement proper role-based access control (RBAC)
12. Azure Container Registry (New)
    - Services: Azure Container Registry
    - Best Practices:
        - Enable container scanning for vulnerabilities
        - Use geo-replication for better performance
        - Implement CI/CD pipelines for automated image builds
        - Use managed identities for pull access
13. Azure Kubernetes Service (New)
    - Services: Azure Kubernetes Service
    - Best Practices:
        - Use node pools for efficient resource allocation
        - Implement proper network policies
        - Use Helm charts for application deployment

    - Enable cluster autoscaling for cost optimization

14. Power BI (New)
    - Services: Power BI
    - Best Practices:
        - Use DirectQuery mode for large datasets
        - Implement row-level security for data privacy
        - Use composite models for complex data relationships
        - Optimize report performance with proper data modeling
15. Azure Synapse Analytics (New)
    - Services: Azure Synapse Analytics
    - Best Practices:
        - Use serverless SQL pool for ad-hoc querying
        - Implement proper data partitioning for better performance
        - Use workload management to balance resource usage
        - Integrate with Azure Machine Learning for advanced analytics

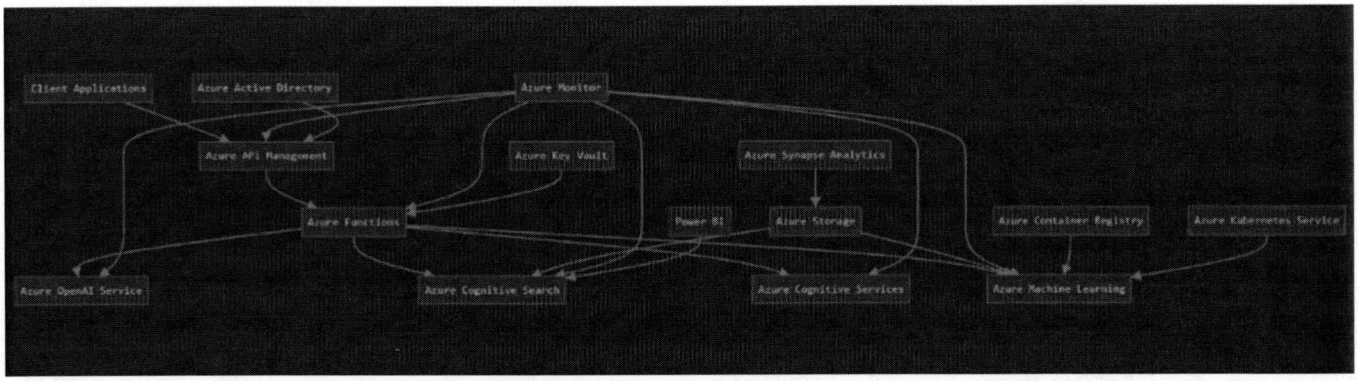

This expanded architecture provides a more comprehensive view of a generative AI solution on Azure. It includes additional services that can enhance data processing, visualization, and deployment capabilities.

When implementing this architecture, consider the following overall best practices:

1. Security: Implement defense in depth, using Azure Security Center, Azure Sentinel, and following the principle of least privilege.
2. Scalability: Design your solution to scale horizontally, using services like Azure Kubernetes Service and Azure Functions.
3. Cost Optimization: Use Azure Cost Management to monitor and optimize your spending.
4. Compliance: Ensure your solution meets relevant compliance standards (e.g., GDPR, HIPAA) using Azure Compliance Manager.
5. DevOps: Implement CI/CD pipelines using Azure DevOps or GitHub Actions for automated testing and deployment.
6. Disaster Recovery: Design for high availability and implement proper backup and disaster recovery strategies.

4. Conclusion

Building Generative AI solutions on involves leveraging a wide range of cloud services, from data processing to deployment and monitoring. By using Cloud provider integrated ecosystem, organizations can implement scalable, secure, and high-performance AI models that address complex business challenges. The reference architecture provided in this chapter serves as a blueprint for designing and deploying state-of-the-art Generative AI models on Cloud, ensuring that all critical aspects—such as data management, model training, and operational monitoring—are covered comprehensively.

Chapter 7

Security Considerations

As organizations increasingly adopt Generative AI (GenAI) to power their applications and business processes, the security implications of deploying these complex models become a critical concern. GenAI models, whether used for natural language generation, image creation, or data synthesis, introduce unique security risks due to their data-intensive nature, potential for misuse, and the sophisticated techniques required to build and deploy them. This chapter explores the key security challenges faced in deploying GenAI on Amazon Web Services (AWS), offers best practices to mitigate these risks, and provides a detailed overview of AWS tools and services designed to secure GenAI environments.

1. Security Challenges in Generative AI Implementations on AWS

The security challenges associated with Generative AI stem from the complexity of the models, the sensitive nature of the data they handle, and the unique threats posed by their ability to generate and manipulate content. Below, we delve into the primary security concerns and implementation challenges for Generative AI on AWS:

1.1 Data Privacy and Confidentiality

Generative AI models are trained on large datasets that may contain sensitive information, such as personally identifiable information (PII), proprietary business data, and intellectual property. Ensuring the confidentiality and privacy of this data during training, storage, and deployment is a top priority.

- **Challenge**: Sensitive data can be exposed during model training, transfer, or storage if not properly secured. This could lead to unintentional data leakage, compliance violations, and potential legal repercussions.
- **Example**: An AI model trained on healthcare data might unintentionally reveal patient details if the training data is not anonymized or if access controls are improperly configured.

Best Practice: Use encryption at rest and in transit, apply strict access controls, and implement data masking and anonymization techniques to protect sensitive information.

1.2. Model Security and Integrity

Generative AI models themselves are valuable assets that need to be protected. Attackers may attempt to access, manipulate, or reverse-engineer models to extract sensitive training data or alter the model's behavior.

- **Challenge**: Unauthorized access to model artifacts, such as training data, model weights, or configurations, can lead to intellectual property theft, model manipulation, or insertion of backdoors.
- **Example**: A competitor or malicious actor might gain access to a model deployed in an unprotected S3 bucket, allowing them to copy the model weights or modify the model's logic.

Best Practice: Store model artifacts in encrypted S3 buckets with strict IAM policies, enable VPC endpoint security, and use AWS Secrets Manager to protect model parameters.

1.3. Identity and Access Management (IAM)

Improper IAM configurations can lead to accidental or malicious access to sensitive resources, such as training data, model endpoints, and monitoring logs.

- **Challenge**: Managing IAM permissions across multiple AWS services (e.g., SageMaker, Lambda, EC2) can become complex, increasing the risk of misconfigured permissions that expose resources.
- **Example**: An overly permissive IAM role might allow a low-privileged user to access and modify model training jobs, leading to data exposure or sabotage.

Best Practice: Implement the principle of least privilege, use role-based access control, and regularly audit IAM roles and policies to minimize attack surfaces.

1.4. Adversarial Attacks

Generative AI models are vulnerable to adversarial attacks, where malicious inputs are crafted to manipulate the model's behavior. These attacks can result in biased, harmful, or misleading outputs.

- **Challenge**: Detecting and defending against adversarial attacks is difficult because these inputs are often subtle and specifically designed to evade standard validation checks.
- **Example**: An attacker might use a carefully crafted input prompt to trick a language model into generating confidential information or producing content that violates compliance policies.

Best Practice: Implement input validation, use adversarial training techniques, and monitor outputs for unusual patterns that might indicate malicious manipulation.

1.5. Compliance and Governance

Generative AI models deployed in regulated industries, such as healthcare, finance, or government, must adhere to strict compliance requirements, including GDPR, HIPAA, and PCI DSS. Ensuring compliance across the entire lifecycle of GenAI models—from data collection to model deployment—is a complex task.

- **Challenge**: Compliance management for Generative AI involves ensuring that data handling, model outputs, and operational processes meet industry-specific regulations, which may vary across regions and applications.

- **Example**: A financial institution deploying a language model for customer support must ensure that the model does not process or store sensitive financial data in ways that violate PCI DSS standards.

Best Practice: Use automated compliance tools, monitor data handling practices, and enforce governance policies across all stages of model development and deployment.

2. Best Practices for Securing Generative AI on AWS

To mitigate the risks associated with Generative AI deployments, organizations should implement a comprehensive set of security best practices. These recommendations address data security, model integrity, IAM configurations, and compliance management.

2.1. Secure Data Management

Data is the backbone of any Generative AI model, making data security a top priority. Implementing strong encryption, access controls, and data anonymization techniques are essential for protecting sensitive information.

- **Encrypt Data at Rest and in Transit**
    - Use AWS Key Management Service (KMS) to manage encryption keys and automate data encryption across services like S3, RDS, and DynamoDB.
    - Ensure that data is encrypted during transfer using Secure Socket Layer (SSL) or Transport Layer Security (TLS).
    - **AWS Tools**: *AWS KMS, Amazon S3 Server-Side Encryption, AWS Transfer Family.*
    - **Implementation Example**: Enable server-side encryption with AWS KMS for an S3 bucket storing training data:

```
{
 "Version": "2012-10-17",
 "Statement": [
  {
   "Sid": "EncryptS3BucketObjects",
   "Effect": "Deny",
   "Principal": "*",
   "Action": "s3:PutObject",
   "Resource": "arn:aws:s3:::my-genai-bucket/*",
   "Condition": {
    "StringNotEquals": {
     "s3:x-amz-server-side-encryption": "aws:kms"
    }
   }
  }
 ]
}
```

- **Implement Data Anonymization and Masking**
  - Use AWS Glue DataBrew for data transformation and anonymization. Mask or remove PII before using datasets for training or testing to ensure compliance and reduce risk.
  - **AWS Tools**: *AWS Glue DataBrew, AWS Lake Formation.*
- **Implement Least Privilege Access for Data Storage**
  - Apply S3 bucket policies and IAM roles to enforce least privilege. Use bucket-level policies to control access to sensitive datasets and ensure that only authorized users can modify or access the data.
  - **AWS Tools**: *Amazon S3 Policies, IAM Roles and Policies, AWS Organizations.*

2.2. Implement Strong Identity and Access Management (IAM)

IAM is at the core of securing resources in AWS. Properly configuring IAM roles, permissions, and access policies ensures that only authorized users can access sensitive resources.

- **Use Fine-Grained IAM Policies**
  - Define IAM policies that grant the least privilege required for a specific role or user. Use service-specific permissions to restrict access to resources such as SageMaker notebooks, EC2 instances, and S3 buckets.
  - **AWS Tools**: *AWS IAM, IAM Access Analyzer, AWS Resource Access Manager.*
- **Enable Multi-Factor Authentication (MFA)**
  - Enforce MFA for all IAM users and roles that access critical resources, such as SageMaker notebooks, EC2 instances, or API gateways exposing model endpoints.
  - **AWS Tools**: *AWS IAM with MFA, AWS SSO.*
- **Use AWS Organizations for Centralized IAM Management**
  - Use AWS Organizations to centrally manage IAM policies, permissions, and billing across multiple AWS accounts. This is especially useful for organizations that have separate environments for development, testing, and production.
  - **AWS Tools**: *AWS Organizations, Service Control Policies (SCPs).*

2.3. Secure Model Deployment and Inferencing

Model deployment is a critical phase in the Generative AI lifecycle that requires robust security measures to protect models from unauthorized access, data leaks, and performance degradation. The following best practices can help secure your models during deployment and inferencing:

- **Deploy Models Within a Secure Virtual Private Cloud (VPC)**

  Deploying models inside a VPC isolates them from the public internet, reducing the risk of unauthorized access. By using Amazon VPC, organizations can control the network configuration and apply security group rules to restrict access to specific IP addresses, subnets, or other AWS services.

  - **AWS Tools**: Amazon VPC, AWS PrivateLink, AWS Direct Connect.
  - **Best Practice**:

    - ✓ Use VPC endpoint policies to control which AWS services and resources can communicate with your models.
    - ✓ Leverage AWS PrivateLink to create private endpoints for your models, allowing secure communication between services within your VPC.
  - o **Implementation Example**: When deploying a SageMaker model, configure it to use a VPC endpoint to prevent external internet access:

```
vpc_config = {
    'SecurityGroupIds': ['sg-123abc45'],
    'Subnets': ['subnet-678def90']
}
sagemaker_model = sagemaker.create_model(
    ModelName='secure-genai-model',
    PrimaryContainer=container,
    ExecutionRoleArn=role,
    VpcConfig=vpc_config
)
```

- **Use HTTPS and SSL/TLS for Secure Communications**

Ensure that all communication between clients and your deployed models is encrypted using HTTPS. SageMaker endpoints should be configured with SSL certificates to encrypt data in transit.

- o **AWS Tools**: Amazon API Gateway, AWS Certificate Manager (ACM), SageMaker Endpoint Configuration.
- o **Best Practice**:
  - ✓ Use Amazon API Gateway as an intermediary between your SageMaker endpoints and external clients, providing built-in support for SSL termination and IAM authentication.
  - ✓ Use AWS Certificate Manager (ACM) to provision and manage SSL/TLS certificates for API Gateway and SageMaker endpoints.
  - **Implementation Example**: Create a secure API Gateway to expose a SageMaker model endpoint:

```
Resources:
  MyApi:
    Type: "AWS::ApiGateway::RestApi"
    Properties:
      Name: "SecureGenAIAPI"
      EndpointConfiguration:
        Types: ["REGIONAL"]
      FailOnWarnings: true
```

- **Implement Access Control Using IAM Policies:** Control access to your SageMaker endpoints using IAM policies that restrict which users, roles, or services can invoke the model. This minimizes the risk of unauthorized access and reduces the attack surface.
  - **AWS Tools**: AWS IAM, SageMaker Endpoint Access Policies, IAM Resource Policies.
  - **Best Practice**:
    - ✓ Define specific IAM policies for each user group, such as data scientists, ML engineers, and DevOps teams, ensuring that each group has access only to the resources they need.
    - ✓ Implement resource-based policies on SageMaker endpoints to restrict access based on conditions like IP addresses, time of day, or AWS accounts.

2.4. Protect Against Adversarial Attacks

Adversarial attacks are a significant threat to Generative AI models, where attackers use crafted inputs to exploit weaknesses in the model's logic, leading to misleading or harmful outputs. Implementing adversarial defenses is crucial to maintaining the integrity and trustworthiness of your models.

- **Implement Input Validation and Sanitization**

  Use input validation to check and sanitize data inputs before they reach your model. This prevents malformed or malicious inputs from affecting the model's behavior.

  - **AWS Tools**: AWS WAF, AWS Shield, AWS Lambda.
  - **Implementation**: Use AWS Lambda as a pre-processing layer to validate and sanitize inputs before sending them to a SageMaker endpoint.
  - **Best Practice**:
    - ✓ Implement rule-based input validation using AWS WAF to detect and block suspicious inputs at the network layer.
    - ✓ Use Lambda functions to filter inputs based on specific criteria (e.g., format, content) before forwarding them to the model.
  - **Implementation Example**: Use AWS WAF to protect a SageMaker endpoint by setting up rules to filter out inputs that contain SQL injection patterns or malformed strings:

```
{
  "Version": "2012-10-17",
  "Statement": [
    {
      "Effect": "Deny",
      "Principal": "*",
      "Action": "s3:PutObject",
      "Resource": "arn:aws:s3:::my-secure-bucket/*",
      "Condition": {
        "StringNotEquals": {
          "s3:x-amz-server-side-encryption": "aws:kms"
        }
      }
    }
  ]
}
```

- **Train Models with Adversarial Robustness Techniques**

Implement adversarial training techniques during model development to improve robustness against adversarial inputs. This involves training the model using adversarial examples to make it less sensitive to minor input manipulations.

  - **AWS Tools**: Custom SageMaker Training Scripts, Amazon SageMaker Debugger.
  - **Best Practice**:
    - ✓ Use adversarial training datasets that include a mix of regular and adversarially crafted inputs to improve model robustness.
    - ✓ Monitor training performance using SageMaker Debugger to identify vulnerabilities during training.

- **Use Differential Privacy to Protect Sensitive Patterns**

Implement differential privacy techniques to obscure sensitive patterns in training data, making it difficult for attackers to extract or reverse-engineer sensitive information from the model.

  - **AWS Tools**: Custom Differential Privacy Implementations, AWS SageMaker Custom Training Jobs.
  - **Best Practice**:
    - ✓ Add noise to training data or model parameters to obscure sensitive patterns.

- ✓ Use secure aggregation techniques to combine model updates without exposing individual data points.

2.5. Ensure Compliance and Governance

Compliance is a critical component of any Generative AI deployment, particularly in regulated industries. AWS provides tools to monitor compliance, enforce governance policies, and maintain auditability across all stages of the GenAI lifecycle.

- **Enable AWS CloudTrail for Comprehensive Logging**

Use AWS CloudTrail to log all API calls and user activities across your AWS environment. This provides a comprehensive audit trail for compliance analysis, forensic investigations, and security monitoring.

  - **AWS Tools**: AWS CloudTrail, Amazon CloudWatch Logs, AWS Config.
  - **Best Practice**:
    - ✓ Enable CloudTrail in all AWS regions and configure it to capture API calls for all AWS services used in your Generative AI solution.
    - ✓ Use CloudWatch to monitor CloudTrail logs for unusual activity patterns.
  - **Implementation Example**: Set up a CloudTrail trail to capture all management events in your AWS account:

```
{
  "trailName": "GenAIAuditTrail",
  "s3BucketName": "genai-logs-bucket",
  "includeGlobalServiceEvents": true,
  "isMultiRegionTrail": true,
  "enableLogFileValidation": true
}
```

- **Use AWS Config for Compliance Monitoring**

AWS Config continuously monitors AWS resources and ensures that they comply with security and governance policies. Config can automatically remediate non-compliant resources, such as unencrypted S3 buckets or publicly accessible EC2 instances.

  - **AWS Tools**: AWS Config, AWS Security Hub.
  - **Best Practice**:
    - ✓ Define Config rules for all compliance requirements, such as encryption, access controls, and network configurations.
    - ✓ Use AWS Security Hub to aggregate compliance findings and monitor your AWS environment from a single dashboard.
  - **Implementation Example**: Create an AWS Config rule to ensure all S3 buckets are encrypted:

```
Resources:
  EncryptedBucketRule:
    Type: "AWS::Config::ConfigRule"
    Properties:
      ConfigRuleName: "s3-bucket-encryption-check"
      Source
Owner: "AWS"
        SourceIdentifier:
"S3_BUCKET_SERVER_SIDE_ENCRYPTION_ENABLED"
```

3. AWS Tools and Services for Securing Generative AI Deployments

AWS offers a comprehensive suite of tools and services specifically designed to secure Generative AI models and their associated infrastructure. Below is a list of key AWS tools and services for building a secure GenAI environment:

- **AWS Identity and Access Management (IAM)**: Manage identities and permissions with fine-grained access control.
- **AWS Key Management Service (KMS)**: Encrypt data and manage encryption keys.
- **Amazon SageMaker**: Deploy models securely, enable monitoring, and manage endpoint access.
- **Amazon VPC**: Isolate models in a private network environment.
- **AWS CloudTrail**: Enable logging and monitoring for all API calls.
- **AWS Config**: Ensure compliance with security and governance policies.
- **AWS Security Hub**: Centralize security findings and monitor compliance.
- **AWS WAF and Shield**: Protect against web-based threats and malicious inputs.

Conclusion

Securing Generative AI deployments on AWS requires a comprehensive strategy that addresses data protection, model integrity, identity management, and compliance. By implementing the best practices outlined in this chapter and leveraging AWS's extensive suite of security tools and services, organizations can build secure, compliant, and resilient Generative AI solutions that protect sensitive data, ensure robust model performance, and maintain compliance with industry regulations

4. Best Practices for Securing Generative AI on Google Cloud

Implementing security best practices across all stages of the Generative AI lifecycle—data preparation, model training, deployment, and operational monitoring—is essential for mitigating risks and ensuring safe operations. Below are the recommended best practices along with relevant Google Cloud tools and code examples.

4.1. Secure Data Management

Data is the foundation of any Generative AI model. Securing sensitive data during storage, transfer, and processing is critical to maintaining confidentiality and compliance.

- **Encrypt Data at Rest and in Transit**

  Use Google Cloud Key Management Service (KMS) to encrypt data at rest in Google Cloud Storage (GCS), BigQuery, and other data services. Enable SSL/TLS for encrypting data in transit.

  - **Google Cloud Tools**: Cloud KMS, Cloud Storage, BigQuery.
  - **Implementation Example**: Create a Cloud KMS key and use it to encrypt data stored in a GCS bucket:

```
# Create a key ring
gcloud kms keyrings create my-key-ring --location=us-central1


# Create a key within the key ring
gcloud kms keys create my-key --location=us-central1 \
  --keyring=my-key-ring --purpose=encryption


# Encrypt a file using the KMS key
gcloud kms encrypt --location=us-central1 --keyring=my-key-ring --key=my-key \
  --plaintext-file=sample.txt --ciphertext-file=sample.txt.enc
```

  - **Best Practice**: Use Customer-Managed Encryption Keys (CMEK) in Cloud KMS for greater control over encryption keys.

- **Implement Data Anonymization and Masking**

  Use Data Loss Prevention (DLP) API to anonymize or mask PII before using it in training datasets. This ensures that sensitive information is not exposed during model training.

  - **Google Cloud Tools**: *DLP API, BigQuery.*
  - **Implementation Example**: Mask email addresses in a dataset using the DLP API:

```
# Initialize DLP client
dlp = dlp_v2.DlpServiceClient()

# Configure de-identification transformation
deidentify_config = {
    'info_type_transformations': {
        'transformations': [
            {'primitive_transformation':   {'replace_with_info_type_config':
{}}}
        ]
    }
}

# Define the item to de-identify
item = {'value': 'Contact: john.doe@example.com'}

# Execute de-identification
response = dlp.deidentify_content(
    parent="projects/my-project",
    deidentify_config=deidentify_config,
    item=item,
)

print("De-identified content: {}".format(response.item.value))
```

- **Best Practice**: Use the DLP API to automatically detect and mask sensitive fields (e.g., SSNs, phone numbers) in datasets used for Generative AI training.

- **Use Fine-Grained Access Control for Data Storage**

Implement IAM policies to enforce least privilege access for data stored in GCS, BigQuery, and other storage services.

- **Google Cloud Tools**: Cloud IAM, Cloud Storage IAM Policies.
- **Implementation Example**: Restrict access to a GCS bucket to only specific users:

```
gsutil iam ch user:alice@example.com:objectViewer gs://my-secure-bucket
```

- **Best Practice**: Use bucket-level IAM policies to restrict access to sensitive datasets.

4.2. Implement Strong Identity and Access Management (IAM)

Managing identity and access permissions is crucial for controlling who can access and manipulate data, models, and other cloud resources.

- **Use Role-Based Access Control (RBAC)**

Define IAM roles that adhere to the principle of least privilege, granting only the permissions necessary for each user role.

  - **Google Cloud Tools**: Cloud IAM, IAM Custom Roles.
  - **Implementation Example**: Create a custom IAM role with specific permissions:

```
gcloud iam roles create CustomDataScientistRole \
  --project=my-project \
  --title="Data Scientist" \
  --permissions=bigquery.tables.get,bigquery.tables.list,storage.objects.get
```

  - **Best Practice**: Use predefined roles for standard use cases and custom roles for specialized permissions.

- **Enable Multi-Factor Authentication (MFA)**

Enforce MFA for all users who access sensitive resources, such as datasets, notebooks, or deployed models.

  - **Google Cloud Tools**: Cloud Identity, Cloud IAM Policies.
  - **Implementation**: Configure MFA using Google Cloud Identity and set up conditional access policies that require MFA for sensitive resources.

- **Use Organization Policies to Enforce Security Standards**

Use organization policies to enforce security best practices across all projects in your Google Cloud organization.

  - **Google Cloud Tools**: Organization Policy Service.
  - **Implementation Example**: Restrict public IP access to Cloud Storage buckets:

```
gcloud resource-manager org-policies enable-enforce \
  --organization=123456789012 \
  --constraint=constraints/compute.vmExternalIpAccess
```

  - **Best Practice**: Use policies to enforce encryption, restrict resource sharing, and prevent public access.

4.3. Secure Model Development and Deployment

Generative AI models, once developed, need to be securely deployed and managed to prevent unauthorized access, tampering, and unintended information leaks. Below are best practices and Google Cloud tools to secure model development and deployment.

- **Use Private and Secure Model Training Environments**

  During the training phase, ensure that models are developed and trained in isolated and secure environments. Use Vertex AI Workbench to run Jupyter notebooks securely within a Virtual Private Cloud (VPC).

  - **Google Cloud Tools**: Vertex AI Workbench, VPC Service Controls.
  - **Implementation Example**: Configure a Vertex AI Workbench notebook instance to run inside a secure VPC environment:

```
gcloud notebooks instances create secure-notebook \
    --vm-image-project=deeplearning-platform-release \
    --vm-image-family=common-cpu \
    --location=us-central1-a \
    --subnet=my-secure-subnet \
    --network=my-vpc-network
```

  - **Best Practice**: Use VPC Service Controls to define security perimeters around Vertex AI, GCS, and other Google Cloud resources to prevent data exfiltration

- **Deploy Models Using Secure Endpoints**

  When deploying models using Vertex AI, ensure that the endpoints are protected with HTTPS and IAM-based access controls. Limit access to model endpoints based on user roles, IP addresses, or other contextual attributes.

  - **Google Cloud Tools**: Vertex AI, Cloud Endpoints, IAM Policies.
  - **Implementation Example**: Deploy a Vertex AI model and restrict access using IAM roles:

```
gcloud beta ai models upload \
    --region=us-central1 \
    --display-name="secure-genai-model" \
    --container-image-uri="gcr.io/my-project/genai-model"
gcloud projects add-iam-policy-binding my-project \
    --member="user:alice@example.com" \
    --role="roles/aiplatform.viewer"
```

   - **Best Practice**: Use Cloud Endpoints with IAM-based authentication to control access to model endpoints.

- **Enable Model Monitoring for Drift and Anomalies**

  Use Vertex AI Model Monitoring to detect data drift, model skew, and anomalies in predictions. This helps ensure that your models perform as expected and that they are not compromised.

   - **Google Cloud Tools**: Vertex AI Model Monitoring, Cloud Monitoring.
   - **Implementation Example**: Set up a monitoring job for a deployed Vertex AI model to track input data for drift:

```
gcloud beta ai model-monitoring-jobs create \
  --model=my-model-id \
  --display-name="model-drift-monitor" \
  --analysis-instance-schema-path=gs://my-bucket/schema.json \
  --target-drift-thresholds=age=0.3,income=0.5
```

   - **Best Practice**: Configure alerts for any deviations in input data patterns or model performance to proactively address security or data quality issues.

- **Use Container Scanning for Deployed Models**

  Before deploying models as containers, use Google Cloud Container Analysis to scan container images for vulnerabilities and potential security issues.

   - **Google Cloud Tools**: Container Analysis, Container Registry.
   - **Implementation Example**: Scan a container image for vulnerabilities:

```
gcloud container images list-tags gcr.io/my-project/my-model-container

gcloud container images describe gcr.io/my-project/my-model-container@sha256:<hash>
```

   - **Best Practice**: Set up automated vulnerability scanning for all container images used in model deployments.

4.4. Protect Against Adversarial Attacks

Generative AI models are vulnerable to adversarial inputs—carefully crafted data designed to manipulate the model's outputs. Defending against these attacks involves using a combination of input validation, adversarial training, and continuous monitoring.

- **Implement Input Validation and Preprocessing**

  Use input validation techniques to filter out malicious inputs before they are processed by the model. This includes validating the format, size, and content of inputs to ensure they meet expected standards.

  - **Google Cloud Tools**: Cloud Functions, Cloud Run, Cloud WAF.
  - **Implementation Example**: Deploy a Cloud Function to validate inputs before passing them to a Vertex AI endpoint:

```
def validate_input(request):
    try:
        data = request.get_json()
        if 'text' not in data or len(data['text']) > 500:
            return "Invalid input", 400
        return data, 200
    except Exception as e:
        return str(e), 400
```

  - **Best Practice**: Use Cloud WAF to filter and block suspicious inputs at the network layer before they reach your model.

- **Train Models Using Adversarial Robustness Techniques**

During the training phase, introduce adversarial examples (inputs designed to mislead the model) to build robustness against these types of attacks. This helps the model learn to ignore or resist malicious inputs.

  - **Google Cloud Tools**: Custom Vertex AI Training Scripts, BigQuery ML for Anomaly Detection.
  - **Implementation Example**: Use a custom Vertex AI training script to include adversarial examples in the training dataset:

```
import tensorflow as tf


def create_adversarial_pattern(model, input_data, label):
    with tf.GradientTape() as tape:
        tape.watch(input_data)
        prediction = model(input_data)
        loss = tf.keras.losses.categorical_crossentropy(label, prediction)
    gradient = tape.gradient(loss, input_data)
```

```
    signed_grad = tf.sign(gradient)
    return signed_grad

# Generate adversarial examples
adversarial_inputs = input_data + 0.1 *
create_adversarial_pattern(model, input_data, label)
```

- **Best Practice**: Use BigQuery ML for anomaly detection to identify and flag inputs that deviate significantly from expected patterns.

- **Use Differential Privacy to Protect Model Outputs**

Implement differential privacy techniques to limit the amount of information that can be extracted from model outputs. This involves adding noise to outputs or using secure aggregation techniques to ensure that sensitive patterns are not exposed.

- **Google Cloud Tools**: Custom TensorFlow Implementations, Vertex AI Secure Aggregation.
- **Best Practice**: Implement secure aggregation in Vertex AI to protect against inference attacks that attempt to extract training data from model outputs.

4.5. Compliance and Governance

Compliance and governance are essential for Generative AI deployments, especially in regulated industries such as finance, healthcare, and government. Google Cloud offers a suite of tools to enforce compliance and maintain governance across your Generative AI projects.

- **Use Google Cloud Armor for Compliance Protection**

Protect your model endpoints with Cloud Armor, which offers security policies for blocking unwanted traffic and mitigating DDoS attacks. Use preconfigured WAF rules to meet compliance requirements.

- **Google Cloud Tools**: Cloud Armor, Cloud Endpoints.
- **Implementation Example**: Configure a Cloud Armor policy to allow traffic only from specific IP addresses:

```
gcloud compute security-policies create secure-endpoint-policy \
    --description="Allow only trusted IPs"
gcloud compute security-policies rules create 1000 \
    --security-policy=secure-endpoint-policy \
    --expression="origin.ip in ['1.2.3.4']" \
    --action=allow
```

- **Enable Audit Logging with Cloud Audit Logs**

Use Cloud Audit Logs to track all activity across your Google Cloud environment. Ensure that all sensitive actions, such as data access or model modifications, are logged and reviewed regularly.

  - **Google Cloud Tools**: Cloud Audit Logs, Cloud Logging.
  - **Implementation Example**: Enable audit logging for Vertex AI to monitor access and modification activities:

```
gcloud projects add-iam-policy-binding my-project \
  --member="serviceAccount:my-service-account@my-project.iam.gserviceaccount.com" \
  --role="roles/logging.logWriter"
```

  - **Best Practice**: Configure log-based alerts for critical actions, such as changes to IAM policies or modifications to sensitive datasets.

- **Use Policy Intelligence to Detect Risky Configurations**

Use Policy Intelligence features like IAM Recommender and Policy Troubleshooter to detect and remediate risky IAM configurations, such as overly permissive roles or unused permissions.

  - **Google Cloud Tools**: IAM Recommender, Policy Troubleshooter.
  - **Implementation Example**: Identify and remove overly permissive roles using IAM Recommender:

```
gcloud recommender recommendations list \
  --project=my-project \
  --location=global \
  --recommender=google.iam.policy.Recommender
```

Conclusion

Securing Generative AI deployments on Google Cloud requires a multi-faceted approach that encompasses data protection, model integrity, access management, compliance, and operational monitoring. By implementing the best practices outlined in this chapter and leveraging Google Cloud's powerful security tools—such as Cloud KMS, Vertex AI, IAM, and Cloud Armor—organizations can build robust, secure, and compliant GenAI solutions that protect sensitive data and ensure trustworthy AI outputs.

5 Best Practices for Securing Generative AI on Azure

Generative AI applications demand stringent security measures to protect data, models, and infrastructure. Azure offers various tools and practices for implementing robust security for Generative AI, ensuring the confidentiality, integrity, and compliance of sensitive information. This chapter explores best practices for securing Generative AI on Azure, covering secure data management, identity and access control, secure model development, adversarial defense, and compliance.

5.1. Secure Data Management

- **Encrypt Data at Rest and in Transit**

  Data encryption is critical for securing sensitive information, both at rest and during transmission. Azure provides robust encryption mechanisms to protect data within storage solutions and while in transit.

- **Encrypting Data at Rest**

  Azure Storage supports encryption at rest by default, using Microsoft-managed keys or customer-managed keys (CMKs) through Azure Key Vault.

```
// ARM Template Snippet for Azure Blob Storage Encryption with Customer-Managed Keys
{
  "name": "SecureBlobStorage",
  "type": "Microsoft.Storage/storageAccounts",
  "apiVersion": "2019-06-01",
  "location": "[resourceGroup().location]",
  "properties": {
    "encryption": {
      "keySource": "Microsoft.Keyvault",
      "keyvaultproperties": {
        "keyvaulturi": "https://<YourKeyVaultName>.vault.azure.net",
        "keyname": "YourKey",
        "keyversion": "<KeyVersion>"
      }
    }
  }
}
```

- **Encrypting Data in Transit**

  Use HTTPS and TLS for secure data transmission. Enforce HTTPS connections for Azure Storage using the following Azure CLI command:

```
az storage account update --name <storage-account-name> --https-only true
```

5.2 Implement Data Anonymization and Masking

For Generative AI models trained on sensitive data, anonymization and masking reduce privacy risks by obfuscating personally identifiable information (PII) before training.

- **Implementing Data Masking with Azure Data Factory**

  Azure Data Factory can be configured to mask or anonymize data fields before data is stored or used for model training.

```
// Azure Data Factory Data Masking Pipeline JSON Snippet
{
  "name": "DataMaskingPipeline",
  "properties": {
    "activities": [
      {
        "name": "MaskPIIData",
        "type": "MappingDataFlow",
        "source": {
          "dataset": "SourceDataset"
        },
        "transformation": [
          {
            "type": "Masking",
            "columns": ["SSN", "Email"],
            "maskingType": "randomize"
          }
        ],
        "sink": {
          "dataset": "MaskedDataset"
        }
      }
    ]
  }
}
```

- **Use Fine-Grained Access Control for Data Storage**

  Azure Role-Based Access Control (RBAC) allows you to assign specific permissions to users or applications, ensuring that only authorized entities access sensitive data.

- ***Setting Up RBAC for Blob Storage***

  The following command grants a specific user read-only access to a storage container:

```
az role assignment create --assignee <user-object-id> --role "Storage Blob Data Reader" --scope
"/subscriptions/<subscription-id>/resourceGroups/<resource-
group>/providers/Microsoft.Storage/storageAccounts/<storage-account-
name>/blobServices/default/containers/<container-name>"
```

5.3 Implement Strong Identity and Access Management (IAM)

Azure Active Directory (Azure AD) provides identity and access management, enforcing strong authentication and fine-grained access control for users and applications.

Enforcing Multi-Factor Authentication (MFA)

MFA requires users to authenticate with multiple factors, adding a layer of security for accessing Generative AI resources.

```
# PowerShell command to enable MFA for a user group

New-AzureADMSConditionalAccessPolicy -DisplayName "Require MFA for Generative AI
Users" -Conditions @{

   Users=@{IncludeGroups=@("<group-id>")}

} -GrantControls @{BuiltInControls=@("mfa")} -State "Enabled"
```

- **Conditional Access Policies**

Conditional Access policies allow you to restrict access based on conditions, such as device compliance or location.

```
# PowerShell command to create a Conditional Access Policy for secure access

$policy = New-AzureADMSConditionalAccessPolicy -DisplayName "Require MFA for Specific Locations" -Conditions @{
    Users=@{IncludeGroups=@("<group-id>")}
    Locations=@{IncludeLocations=@("AllTrustedLocations")}
} -GrantControls @{BuiltInControls=@("mfa")} -State "Enabled"
```

5.4. Secure Model Development and Deployment

Ensuring security throughout the model development and deployment lifecycle involves encryption, access control, and monitoring of the Azure Machine Learning environment.

- **Model Encryption**

Encrypt Azure Machine Learning model artifacts using customer-managed keys (CMKs) stored in Azure Key Vault.

```
# YAML snippet for configuring Azure ML workspace encryption with CMK
properties:
  encryption:
    keyVaultProperties:
      keyVaultId: "/subscriptions/<subscription-id>/resourceGroups/<resource-group>/providers/Microsoft.KeyVault/vaults/<key-vault-name>"
      keyIdentifier: "https://<key-vault-name>.vault.azure.net/keys/<key-name>"
```

- **Secure Model Deployment with Private Link**

Deploy models with Azure Private Link to restrict access, preventing exposure to the public internet.

```
// ARM Template snippet to create a Private Endpoint for Azure Machine Learning
{
  "name": "PrivateEndpointForAML",
  "type": "Microsoft.Network/privateEndpoints",
  "properties": {
    "privateLinkServiceConnections": [
      {
        "name": "mlPrivateLink",
        "privateLinkServiceId": "/subscriptions/<subscription-id>/resourceGroups/<resource-group>/providers/Microsoft.MachineLearningServices/workspaces/<workspace-name>",
        "groupIds": ["amlworkspace"]
      }
    ]
  }
}
```

5.5 Protect Against Adversarial Attacks

Adversarial attacks can manipulate models by introducing malicious inputs. Azure provides tools to secure models against such attacks.

Implementing Adversarial Training: Train models with adversarial examples to enhance resilience. Use the Azure Machine Learning SDK to implement adversarial defenses in your training pipeline.

```
from azureml.core import Workspace, Experiment
from azureml.train.sklearn import SKLearn

# Set up the workspace and experiment
ws = Workspace.from_config()
experiment = Experiment(workspace=ws, name='AdversarialTraining')

# Configure and run adversarial training
config = SKLearn(source_directory='.', compute_target='cpu-cluster',
entry_script='train_with_adversarial.py')
run = experiment.submit(config)
```

- **Monitoring for Adversarial Activity**

  Enable logging in Azure Monitor and set alerts for anomalous activity, which may indicate adversarial interference.

```
# PowerShell command to create an alert in Azure Monitor

New-AzMetricAlertRule -Name "AdversarialActivityAlert" -ResourceGroupName <resource-group> -TargetResourceId "/subscriptions/<subscription-id>/resourceGroups/<resource-group>/providers/Microsoft.MachineLearningServices/workspaces/<workspace-name>" -MetricName "RequestCount" -Operator GreaterThan -Threshold 100 -TimeAggregation "Total" -WindowSize "PT5M"
```

5.6 Compliance and Governance

Azure provides several services to ensure compliance with regulations such as GDPR, HIPAA, and others. Use Azure Policy, Azure Security Center, and Azure Blueprints for continuous compliance monitoring.

Implementing Compliance Policies with Azure Policy: Azure Policy enforces compliance policies across resources, detecting and correcting violations in real time.

```
// Azure Policy Definition to enforce encryption
{
  "properties": {
    "displayName": "Enforce Storage Encryption",
    "policyType": "BuiltIn",
    "mode": "Indexed",
    "parameters": {},
    "policyRule": {
      "if": {
        "field": "type",
        "equals": "Microsoft.Storage/storageAccounts"
      },
      "then": {
        "effect": "auditIfNotExists",
        "details": {
          "type":
"Microsoft.Storage/storageAccounts/encryption.keySource",
          "exists": false
        }
      }
    }
  }
}
```

- **Governance with Azure Blueprints**

Azure Blueprints enable repeatable governance and compliance setups, helping organizations quickly deploy policies and security configurations.

```
# PowerShell to create a new Blueprint for secure AI workloads

New-AzBlueprint -Name "GenerativeAIComplianceBlueprint" -Description
"Compliance and security blueprint for Generative AI on Azure" -
ResourceGroupName "<resource-group>"
```

Conclusion

Securing Generative AI implementations on Azure requires a comprehensive approach that encompasses data protection, identity management, model security, and compliance. By following these best practices, organizations can mitigate security risks, ensure regulatory compliance, and deploy Generative AI solutions that are robust, scalable, and secure. Azure's extensive suite of tools, including Azure AD, Key Vault, and Policy, offers a secure foundation for managing sensitive data, protecting AI models, and maintaining stringent security standards.

Chapter 8

Prompt Engineering in AI

Prompt engineering has emerged as a critical skill in the age of advanced artificial intelligence, particularly with the rise of large language models (LLMs) like OpenAI's GPT series, Google's PaLM, and Meta's LLaMA models. While traditional programming involves providing step-by-step instructions to a computer, prompt engineering involves designing specific inputs—or *prompts*—that guide the AI to produce desired outputs. This chapter delves deeply into the concept, techniques, best practices, and real-world applications of prompt engineering, highlighting its growing importance in harnessing the power of modern AI.

1. Understanding Prompt Engineering

Prompt engineering is the practice of crafting and refining the textual inputs that guide a generative AI model's responses. The process resembles a form of linguistic programming, where the primary task is not to write code, but to design *natural language prompts* that instruct the AI on what kind of output is desired. Because the quality and structure of a prompt can have a significant impact on the AI's responses, effective prompt engineering requires a nuanced understanding of language and model behavior.

1.1 Key Concepts in Prompt Engineering

Prompt: A prompt is the textual input that is given to an AI model to generate a response. It could be a question, a statement, a piece of context, or a complex series of instructions.

Example: If you ask, *"Write a short story about a robot discovering emotions,"* the AI will generate a narrative based on the structure and content of your prompt.

Instructional Prompts: These prompts provide direct instructions to the model. For instance, *"Summarize this document in bullet points"* or *"List the top five advantages of renewable energy."* Instructional prompts are explicit and guide the model to focus on a particular type of output.

Contextual Prompts: These include additional context that helps the model understand the scenario. For example, *"You are an expert in 19th-century history. Explain the impact of the Industrial Revolution."* By providing context, the model is more likely to produce an informed and tailored response.

Role-Based Prompts: These prompts assign a role to the AI to influence the style and content of its response. For example, *"Act as a software engineer and explain how to implement a binary search algorithm."* This helps the model frame its response from a specific perspective.

1.2 Why Prompt Engineering Matters

Prompt engineering is essential because of the *context-dependent* nature of LLMs. Large language models operate based on the context they are given, generating outputs that are probabilistically influenced by the input text. Thus, even minor changes in wording can dramatically alter the output, making prompt engineering both a challenge and an opportunity. This means that for a given problem, different prompts might produce different results in terms of quality, coherence, and relevance.

1.3 Best Practices

Use Clear and Specific Instructions: Ensure that the prompt is as clear as possible. Ambiguity in prompts can lead to undesired outputs. For instance, instead of "Describe the benefits of AI," a more specific prompt would be "Describe three specific benefits of AI in the healthcare industry, focusing on patient diagnosis, drug discovery, and treatment planning."

Include Relevant Context: If the prompt requires background knowledge, include it in the input. For example, if asking for a financial summary, provide some initial details: "Given that the company has seen a 10% increase in sales over the last quarter, summarize the potential implications for its annual revenue."

Set Constraints: Adding length or style constraints can prevent verbose or off-topic outputs. For example, "Write a 150-word summary of the article" or "Respond in a formal tone suitable for a business email."

2. How Large Language Models Understand Prompts

Large language models like GPT-4 are built on complex architectures called *Transformers*, which use layers of attention mechanisms to process and generate text. They are trained on vast amounts of data, allowing them to learn intricate patterns in language and context. When a prompt is given, the model generates the most probable continuation of that text based on what it has learned.

2.1 How the Model Interprets a Prompt

The model processes a prompt by breaking it into smaller units called *tokens*. Each word, phrase, or character in the prompt is tokenized, and the model uses these tokens to predict the next sequence of text. It evaluates the prompt based on patterns, syntax, and context, considering both immediate and long-range dependencies.

Example:

Prompt: *"Translate the following English sentence into French: The quick brown fox jumps over the lazy dog."*

The model understands this prompt by recognizing the task (translation) and the target language (French). It then translates each segment of the sentence, considering grammatical rules and context.

2.2 Techniques to Improve Model Understanding

Use Prompt Templates: Templates are predefined prompt structures designed to maintain consistency across different tasks. For example, a template for summarizing articles might look like: "Summarize the following article in 100 words. Focus on key arguments and conclusions: [Text]."

Prompt Length and Token Management: Language models have a maximum token limit (e.g., 4096 tokens in GPT-3). When creating long prompts, prioritize key information to ensure the model receives the most relevant context.

Leveraging Attention Mechanisms: Using structured prompts can help guide the model's attention. For example, sectioning the prompt into *"Introduction"*, *"Body"*, and *"Conclusion"* can help the model understand the flow and produce a coherent response.

3. Techniques for Effective Prompt Engineering

Creating effective prompts is both an art and a science. Prompt engineering is a process of iterative refinement—starting with a simple prompt, analyzing the output, and gradually improving the input to optimize the AI's performance.

3.1 Core Techniques

1. Clarity and Specificity: Ensure that the prompt is explicit and clear. Vague instructions can confuse the model, leading to incomplete or irrelevant responses.

Example:

Weak Prompt: *"Summarize the article."*

Strong Prompt: *"Summarize the article in three bullet points, focusing on the main arguments and supporting evidence."*

2. Using Examples: Providing one or more examples (few-shot prompting) helps the model understand the desired output format. This is especially useful for complex or nuanced tasks.

Example:

Prompt: *"Here is a list of fruits: 1. Apple, 2. Banana, 3. Mango. Continue the list with two more fruits."*

The model is more likely to produce a response that follows the exact pattern.

3. Embedding Constraints: Adding explicit constraints (length, style, tone) can refine the model's outputs.

Example:

Prompt: *"Generate a 50-word product description for a smartphone, highlighting its camera features and battery life. Use a professional tone."*

4. Role-Playing Prompts: Assigning roles can guide the model's perspective and style.

Example:

"You are a health consultant. Advise a client on the benefits of a balanced diet and regular exercise."

5. Prompt Chaining and Multi-Step Prompts: For complex tasks, break down the problem into smaller, sequential steps.

Example:

Step 1: *"Identify the main topic of the document."*

Step 2: *"Summarize the key points related to the main topic."*

Step 3: *"Provide a counterargument to the main points."*

By chaining prompts together, the model can build on its previous responses.

4. Practical Applications of Prompt Engineering

Prompt engineering is used across industries, enabling tailored solutions for a variety of tasks. Some key applications include:

4.1 Content Creation and Copywriting

Prompt engineering is used to create compelling marketing content, product descriptions, and even full-length articles.

Example:

Prompt: *"Write a 200-word product description for a new smartwatch, emphasizing its fitness tracking features and water resistance. Include a call-to-action encouraging customers to learn more."*

4.2 Data Analysis and Summarization

AI models can be prompted to extract insights from large datasets or summarize complex documents.

Example:

Prompt: *"Analyze the following customer reviews for sentiment. Classify each review as positive, neutral, or negative, and provide a short justification for your classification."*

5. Challenges in Prompt Engineering

While powerful, prompt engineering has its limitations:

Ambiguity: Even subtle ambiguities in language can lead to significant misinterpretations by the model.

Biases and Ethical Concerns: Prompts can unintentionally reinforce biases present in training data, resulting in skewed or harmful outputs.

Reproducibility: Small variations in prompts can yield different results, making it difficult to standardize outputs.

6.0 Integrating Prompt Engineering into the GenAI Reference Architecture

Prompt engineering is an essential element of the Generative AI (GenAI) ecosystem, serving as the interface through which users and applications interact with complex AI models to produce desired outputs. In a GenAI reference architecture, prompt engineering acts as the bridge between *model capabilities* and *user requirements*. This makes it particularly relevant for scenarios where the AI models need to understand specific tasks, respond accurately to user queries, or generate precise content. By effectively designing prompts, organizations can tailor AI outputs to meet their goals, improve model performance, and minimize undesired responses.

This section dives deep into how prompt engineering fits into various stages of a GenAI reference architecture, elaborating its role and use cases across model development, deployment, monitoring, and integration layers. We also explore a practical example in a customer service scenario to show how prompt engineering can be applied in real-world contexts.

6.1. Prompt Engineering in the Model Development and Training Phase

During the *Model Development and Training* phase, prompt engineering helps refine and optimize the interactions between the AI model and its users, shaping how the model interprets tasks and outputs relevant results. In this stage, prompt engineering can be leveraged to:

Create Training Data Using Prompts

Creating high-quality training data is essential for developing models that respond effectively to prompts. Prompt engineers can design diverse prompts to generate synthetic training datasets, covering a wide range of scenarios and edge cases.

Example: Suppose a company is training a model to handle customer service inquiries. The prompt engineers might create varied scenarios like:

"Generate a response to a customer complaint about a delayed order, apologizing and offering a discount."

"Respond to a customer asking about return policies for electronic items."

These prompts are used to generate synthetic dialogues, which are then used to fine-tune the model, helping it understand customer service contexts better.

Optimize Model Performance

Prompt engineers work closely with data scientists and AI developers to optimize the prompts used during training, ensuring that the AI learns the desired behaviors. This includes:

Designing *instructional prompts* that specify the format, tone, and style of responses.

Experimenting with different prompt structures (e.g., varying the sequence of instructions) to see which configurations yield the most coherent and contextually appropriate outputs.

Example Integration: In a financial advisory application, prompt engineers can design prompts like *"Provide a brief summary of this quarterly report focusing on net income and cash flow changes. Use professional language suitable for an investor audience."* This guides the AI to extract specific financial information and format it for a professional audience.

Fine-Tuning Models with Instruction-Based Training

Models like OpenAI's InstructGPT are trained using prompts that define specific instructions. Prompt engineers play a key role in crafting these instructions and creating structured datasets that are used to fine-tune the model's behavior.

Example Prompt: *"Given a customer review that contains mixed feedback, identify the positive and negative aspects and present them in separate bullet points."*

This type of prompt is used repeatedly during training, so the model learns to understand complex instructions and respond accurately.

6.2 Prompt Engineering in the Model Deployment and Serving Phase

Once models are trained, they need to be deployed in environments where users can interact with them. In the *Model Deployment and Serving* phase, prompt engineering ensures that the deployed models respond accurately and contextually to real-world inputs. At this stage:

User Interaction Layer: Managing Prompts in Real-Time Scenarios

In a GenAI architecture, the *user interaction layer* acts as the entry point where prompts are fed into the system. Effectively managing prompts at this layer ensures that user inputs are interpreted correctly and produce the intended outputs.

Role of Prompt APIs: Prompt engineering often involves designing and managing *prompt APIs* that serve as standardized interfaces for sending user inputs to the AI model. These APIs define how prompts are structured, what inputs are required, and the expected output format.

Example Integration: In a customer service chatbot, prompt engineers might define templates for common inquiries such as *"Where is my order?"*, *"What is your return policy?"*, and *"How do I reset my password?"* Each of these prompts is mapped to a predefined template that ensures the AI understands the query and responds appropriately.

Using Prompt Templates and Prompt Chaining

Prompt engineers design *prompt templates* and *prompt chaining* structures to maintain consistency and reliability in AI outputs. Prompt chaining is particularly useful for handling complex tasks that require multiple steps.

Prompt Templates: These are predefined prompts with placeholders that can be filled dynamically based on user inputs. For instance, a prompt template for summarizing legal documents might look like:

"Summarize the following legal document: [Document Content]. Focus on identifying key clauses related to compliance and risk management."

Prompt Chaining: For multi-step interactions, the output of one prompt is used as the input for the next. This technique is crucial in customer service scenarios where multiple pieces of information are needed before a final response is given.

Example: In a customer service workflow:

"Identify the issue from the customer's message."

"Retrieve the relevant policy or solution based on the identified issue."

"Draft a personalized response that addresses the issue and references the solution."

6.3. Prompt Engineering in the Model Monitoring and Optimization Phase

After models are deployed, they require continuous monitoring and optimization to ensure they perform as expected in production environments. Prompt engineering plays a role in identifying areas of improvement and refining the prompts to enhance performance.

Improving Model Accuracy Through Prompt Refinement

Prompt engineers analyze model responses to identify patterns of misinterpretation or incorrect outputs. By tweaking the prompt structure, adding constraints, or rephrasing instructions, the AI's accuracy can be significantly improved.

Example: If a customer service model frequently misinterprets queries about "order cancellations" as "order tracking," prompt engineers can refine the prompt to include more specific instructions, such as *"If the query mentions 'cancel' or 'refund,' prioritize providing cancellation information."*

Prompt Evaluation Tools

Prompt evaluation tools, such as OpenAI's Prompt Designer, enable engineers to test different variations of prompts and compare outputs. This helps in understanding how minor changes to prompts influence AI behavior and in choosing the most effective prompt configurations.

Bias and Ethics Considerations

Prompt engineering is crucial in mitigating biases in AI responses. By carefully crafting prompts that exclude biased language or provide balanced perspectives, prompt engineers can reduce the risk of generating harmful or biased content.

Example: In a scenario where an AI is used for recruitment, prompts can be engineered to avoid gender or racial biases by explicitly instructing the model to focus on skills and experience rather than subjective criteria.

6.4. Prompt Engineering in Business Application Integration

In a comprehensive GenAI reference architecture, *Business Applications* represent the layer where AI outputs are consumed and utilized for real-world decision-making. Prompt engineering is vital in tailoring AI outputs to meet the needs of various business units, ensuring that the AI delivers value and aligns with business goals.

Customizing Outputs for Different Stakeholders

Different departments may require tailored outputs from the same generative model. For instance, marketing teams may need creative copy, while product teams need technical descriptions. Using prompt engineering, one can create prompt variations that guide the model to produce distinct outputs for various use cases.

Example Integration: In a financial institution, the marketing team may use a prompt like, *"Generate an engaging blog post on the benefits of our new mortgage plan,"* while the legal team might use, *"Generate a detailed policy document explaining the terms and conditions of the mortgage plan."* Both prompts are directed at the same model but are crafted to produce outputs suited for different stakeholders.

Creating Multi-Step Workflows Using Prompt Chaining

Prompt chaining allows complex business processes to be broken down into a series of smaller, interconnected tasks. Each step is guided by a specific prompt, and the output of one step is used as the input for the next.

Example Integration: In a healthcare setting, the workflow for patient support might look like this:

"Extract key symptoms mentioned by the patient."

"Based on the symptoms, suggest potential conditions to investigate."

"Generate a personalized report for the patient summarizing the findings and recommending further steps."

By designing prompts for each step, prompt engineers ensure that the AI system maintains context and accuracy throughout the workflow.

6.5. Role of Prompt Engineering in Model Security and Compliance

Prompt engineering also plays a role in ensuring that models adhere to security and compliance standards. By designing prompts that enforce security rules and validate compliance requirements, prompt engineers can help organizations maintain regulatory standards and avoid unintended consequences.

Prompt Constraints for Data Privacy

Prompt engineers can include constraints in the prompts that prevent the AI from processing sensitive data or trigger warnings when such data is detected. This is crucial for maintaining data privacy, especially in regulated industries like healthcare and finance.

Example Integration: A prompt for a financial advisor AI might be designed as: *"If the input contains client account numbers or other sensitive identifiers, do not process the input and return the following error message: 'Sensitive data detected. Please redact the data and try again.'"*

Ensuring Compliance in Outputs

Prompt engineers can design prompts to ensure that the AI's outputs comply with industry standards. For example, legal documents generated by AI can be prompted to include specific compliance language, ensuring that the final output meets regulatory guidelines.

Example Integration: In a legal advisory application, prompts can be designed to ensure that AI-generated contracts include mandatory clauses and disclaimers, such as, *"Always include a compliance clause in Section 4 of the contract and label it as 'Regulatory Compliance Obligations.'"*

7. AWS Tools for Prompt Engineering

Amazon Web Services (AWS) provides a comprehensive set of cloud-based tools and services that support the development, deployment, and optimization of AI models, including the crucial task of prompt engineering. As prompt engineering involves crafting precise inputs for generative models to achieve desired outputs, AWS offers specialized resources and infrastructure to streamline this process, making it easier to design, test, and refine prompts. This section explores how AWS tools can be effectively leveraged to support prompt engineering in a generative AI architecture.

7.1. AWS SageMaker for Model Training and Prompt Optimization

Amazon SageMaker is AWS's fully managed service for building, training, and deploying machine learning models. It includes several features that directly assist with prompt engineering, enabling data scientists and developers to optimize the way prompts are designed and used with language models.

SageMaker Studio for Experimentation and Testing

AWS SageMaker Studio is an integrated development environment (IDE) specifically designed for machine learning. It offers a collaborative environment where prompt engineers can experiment with different prompts, test their effectiveness, and refine them to improve the AI model's outputs.

Use Case: Prompt engineers can use SageMaker Studio notebooks to interact with a large language model, design and run prompt variations, and compare outputs in real time. This environment allows for rapid iteration, making it ideal for refining prompts to handle specific use cases, such as summarization, translation, or question-answering.

Feature Highlight: SageMaker Studio's *experiments tracking* feature allows engineers to log different prompt configurations and measure their impact on output quality, which is crucial for understanding the nuances of prompt behavior.

SageMaker JumpStart for Pre-Built Models and Templates

SageMaker JumpStart provides access to a library of pre-trained models and templates for various NLP tasks, such as text classification, summarization, and language generation. These models can be fine-tuned with custom prompts for specific business needs.

Use Case: A prompt engineer can start with a pre-trained language model from SageMaker JumpStart and use it to design prompts tailored to a healthcare chatbot, e.g., *"Summarize patient symptoms and suggest the next steps based on severity."* By using pre-built models, engineers can focus on prompt optimization rather than starting from scratch.

Feature Highlight: JumpStart includes *fine-tuning templates*, which can be modified with custom prompts to quickly adapt the model to different scenarios, making it easier to implement and test prompt variations.

SageMaker Pipelines for Prompt Fine-Tuning and Optimization

AWS SageMaker Pipelines automates the process of model building and evaluation, enabling prompt engineers to integrate prompt testing into the ML workflow. This automation is beneficial when conducting A/B testing on different prompts to evaluate their effectiveness.

Use Case: Engineers can create pipelines that automatically test multiple prompt configurations with a fine-tuned model, compare their outputs, and select the optimal prompts for deployment.

Feature Highlight: Pipelines allow for *automated prompt evaluation*, reducing the time and effort needed to identify the best-performing prompts for specific use cases.

7.2. Amazon Comprehend for NLP and Contextual Understanding

Amazon Comprehend is a natural language processing (NLP) service that uses machine learning to analyze and understand text. It provides features like entity recognition, sentiment analysis, keyphrase extraction, and language detection, which are useful for creating contextually aware prompts.

Creating Context-Aware Prompts

Amazon Comprehend can analyze a body of text to identify key entities, topics, and sentiment, which can be used to craft context-aware prompts for large language models.

Use Case: Before designing a prompt for a generative model, a healthcare application might use Comprehend to analyze patient records and identify key medical terms, diagnoses, and medications. This context can then be embedded into prompts, such as: *"Based on the patient's history of hypertension and current medication regimen, suggest possible side effects that may require monitoring."*

Feature Highlight: By leveraging entity recognition and topic modeling, Comprehend helps engineers create detailed, context-specific prompts that guide the model to generate relevant and precise outputs.

Sentiment-Based Prompt Adjustments

Comprehend's sentiment analysis feature can be used to evaluate the tone and emotional content of prompts, ensuring that the AI generates responses that are appropriate for sensitive scenarios.

Use Case: In customer service applications, Comprehend can help evaluate whether a prompt like *"Apologize for the delay and offer a discount"* conveys the right sentiment for a particular customer's tone.

Feature Highlight: Sentiment analysis can be integrated into prompt refinement workflows, helping prompt engineers tweak wording to achieve the desired emotional impact in generated responses.

7.3. Amazon Lex for Conversational AI and Chatbots

Amazon Lex is a service for building conversational interfaces using voice and text. Lex supports the development of complex conversational bots, which are often built using prompt engineering to guide the bot's responses and maintain context throughout interactions.

Using Intent-Based Prompts for Dialogues

Amazon Lex's core concept revolves around *intents*, which define the purpose of a user's input. Prompt engineers can leverage these intents to structure conversations, ensuring that each prompt elicits the appropriate response from the model.

Use Case: For a healthcare chatbot, an intent like *"GetPatientSymptoms"* could trigger a series of prompts designed to collect comprehensive symptom data:

"Can you describe your symptoms in detail?"

"How long have you been experiencing these symptoms?"

"Have you noticed any other unusual changes in your health recently?"

This structured approach ensures that the AI collects all relevant information before making suggestions or escalating to a human clinician.

Designing Contextual Prompts for Conversational Continuity

Lex can maintain session attributes, which helps preserve context across multiple turns in a conversation. Prompt engineers can use these attributes to design follow-up prompts that build on previous responses.

Use Case: If a patient mentions, *"I've been feeling dizzy and have a headache,"* the follow-up prompt could automatically adjust to include relevant context:

"Thank you for sharing that. You mentioned dizziness and a headache. Can you tell me if you've also experienced nausea or vision changes?"

Feature Highlight: Lex's ability to manage context allows prompt engineers to create dynamic, contextually aware prompts that improve the naturalness and accuracy of conversations.

7.4. AWS Lambda for Serverless Prompt Execution and Integration

AWS Lambda is a serverless compute service that lets you run code without provisioning servers. Lambda can be used to execute and manage prompts programmatically, enabling dynamic prompt generation and integration with other AWS services.

Dynamic Prompt Generation with AWS Lambda

Lambda functions can dynamically generate prompts based on real-time data, user inputs, or external APIs, making it possible to adapt prompts based on changing contexts.

Use Case: In a financial advisory application, a Lambda function can be triggered to generate a prompt like: *"Provide a market analysis based on today's stock performance and recent economic news."* The prompt is generated using real-time data pulled from AWS Data APIs.

Integrating Multiple AWS Services for Complex Prompt Workflows

Lambda can serve as the orchestration layer in a complex GenAI architecture, integrating Amazon Comprehend, SageMaker, and Lex. For example, a Lambda function could:

Use Comprehend to analyze a patient's message for entities and sentiment.

Send the extracted entities to a SageMaker model to generate a preliminary diagnosis.

Pass the output to Lex to craft an appropriate follow-up prompt.

This workflow demonstrates how Lambda can facilitate prompt engineering by automating complex, multi-step interactions.

7.5. Amazon Polly and AWS Translate for Multi-Language Prompt Engineering

Amazon Polly and AWS Translate enable multi-language support for generative models, expanding the reach of prompt engineering across different languages and regions.

Using AWS Translate for Multi-Language Prompts

AWS Translate can convert English prompts into multiple languages, making it easier to localize AI models for different regions.

Use Case: A healthcare application can use Translate to convert an English prompt like, *"Describe your symptoms in detail, including any pain or discomfort,"* into Spanish: *"Describa sus síntomas en detalle, incluyendo cualquier dolor o malestar."*

Amazon Polly for Voice-Based Prompts

Amazon Polly converts text into lifelike speech, enabling voice-based interactions with generative models. This is especially useful in healthcare applications where accessibility is a concern.

Use Case: A virtual assistant can use Polly to read out a dynamically generated prompt: *"It sounds like you're experiencing some discomfort. Would you like me to connect you with a healthcare professional for further assistance?"*

7.6. Amazon CloudWatch for Monitoring and Logging Prompt Effectiveness

Amazon CloudWatch provides monitoring and observability for AWS resources. It can be used to track the performance of prompts, measure response quality, and identify potential issues.

Using CloudWatch for Prompt Performance Monitoring

CloudWatch can log prompt-response pairs and evaluate metrics like response time, accuracy, and sentiment, providing valuable insights into how well prompts are performing.

Use Case: In a customer service chatbot, CloudWatch can track the effectiveness of a prompt like, *"How can I help you today?"* by logging the number of successful resolutions following the prompt.

Feature Highlight: CloudWatch metrics can be used to trigger automatic adjustments to prompts if certain thresholds are met, such as a high rate of ambiguous responses.

8. Google Cloud Tools for Prompt Engineering

Google Cloud offers a rich suite of tools and services that enable prompt engineering for Generative AI applications. These tools empower developers, data scientists, and prompt engineers to build, test, and optimize prompts for large language models (LLMs) and other AI-driven systems. With Google's cutting-edge technologies such as Vertex AI, Dialogflow, and Cloud Natural Language API, prompt engineering becomes a structured and scalable process, providing flexibility for designing advanced AI solutions in areas like customer service, content generation, healthcare, and beyond.

This section explores how Google Cloud tools can be leveraged to support prompt engineering within a Generative AI architecture, covering model development, deployment, monitoring, and real-world applications.

8.1. Vertex AI: Comprehensive Platform for Model Development and Prompt Engineering

Vertex AI is Google Cloud's unified machine learning (ML) platform, designed to build, train, deploy, and manage ML models. It is particularly valuable for Generative AI applications, offering tools for creating and fine-tuning models, managing data, and integrating prompt engineering workflows.

Vertex AI Workbench for Interactive Prompt Development

Vertex AI Workbench provides a collaborative environment with managed Jupyter notebooks, allowing data scientists and prompt engineers to experiment with different prompt structures, test outputs, and refine interactions with large language models.

Use Case: Engineers can use Vertex AI Workbench to test various prompts for a customer service chatbot, such as:

"Respond to a customer complaint about a delayed delivery in a polite and apologetic tone."

"Explain the benefits of our loyalty program to a new customer in less than 100 words."

The workbench environment supports interactive testing, making it easy to evaluate prompt effectiveness and refine prompts based on model responses.

Feature Highlight: With Vertex AI Workbench, engineers can log different versions of prompts, analyze outputs, and integrate testing workflows into a broader machine learning pipeline.

Vertex AI Pipelines for Prompt Fine-Tuning and Automation

Vertex AI Pipelines automate the process of model training, evaluation, and deployment, making it easier to include prompt engineering steps in an automated ML workflow.

Use Case: Engineers can design pipelines that test multiple prompt configurations, track performance metrics (e.g., accuracy, response quality), and automatically select the best-performing prompts. This is particularly useful for large-scale applications where fine-tuning hundreds of prompts manually would be infeasible.

Feature Highlight: Vertex AI Pipelines allow for complex, multi-step workflows that include data preprocessing, prompt testing, and output evaluation, ensuring a streamlined and repeatable prompt optimization process.

Vertex AI Model Monitoring for Prompt Effectiveness

After deploying a generative model, prompt engineers can use Vertex AI Model Monitoring to track the performance of different prompts in real-world interactions. This includes detecting model drift, output quality degradation, or unexpected behaviors.

Use Case: In a healthcare application, Vertex AI Model Monitoring can track how prompts like, *"Describe the side effects of a new medication,"* perform over time. If the model starts generating less accurate descriptions, prompt engineers are alerted to update the prompts or retrain the model with new data.

Feature Highlight: Engineers can set up custom monitoring rules to detect specific issues with prompts, such as a sudden increase in irrelevant responses, and trigger automatic retraining or refinement processes.

8.2. Dialogflow CX: Advanced Conversational AI for Dialogue Management

Dialogflow CX is Google's conversational AI platform, designed for building complex, multi-turn dialogue systems. It provides an ideal environment for prompt engineering in chatbot and voice assistant applications, allowing developers to create, test, and deploy conversational agents that interact seamlessly with users.

Designing Intent-Based Prompts for Dialogue Agents

Dialogflow CX uses *intents* to define user goals and *training phrases* to recognize user inputs. Prompt engineers can design structured intents and corresponding prompts to guide the conversational flow.

Use Case: For a customer support bot, prompt engineers can define an intent for handling refund requests:

Intent: *"RefundRequest"*

Prompt Example: *"I'm sorry to hear you're considering a refund. Could you tell me a bit more about the issue you faced with your purchase?"*

This prompt helps the bot gather relevant information and display empathy, improving user satisfaction.

Feature Highlight: Dialogflow's *flow builder* allows engineers to visualize conversation flows, making it easy to design and refine multi-turn dialogues that adapt to user inputs.

Using Dialogflow CX for Context Management and Follow-Up Prompts

Dialogflow CX's context management capabilities enable prompt engineers to maintain conversation state across multiple interactions, ensuring that the AI model responds accurately even when users switch topics or revisit previous questions.

Use Case: In a healthcare chatbot, if a user mentions, *"I'm feeling nauseous,"* the context is set to "symptom_reporting." Follow-up prompts can then be structured to explore related symptoms:

"I see you mentioned nausea. Are you also experiencing dizziness or headaches?"

The context management ensures that the bot remembers the previous conversation and provides contextually relevant prompts.

Feature Highlight: Engineers can set up *contextual prompts* that adjust dynamically based on user responses, allowing for more natural and coherent interactions.

Custom Payloads and Rich Responses

Dialogflow CX supports custom payloads, allowing prompt engineers to craft complex, rich responses that include images, buttons, and quick replies, enhancing user engagement.

Use Case: In an e-commerce scenario, if a user asks about available colors for a product, the response could include a prompt like:

"The item is available in these colors:" followed by buttons labeled with color options, making it easier for the user to select an option.

8.3. Cloud Natural Language API for Text Analysis and Contextual Prompt Design

Google Cloud's Natural Language API provides powerful text analysis capabilities, including entity recognition, sentiment analysis, and syntax analysis. These features support prompt engineers by providing insights that can be used to design more effective and contextually appropriate prompts.

Entity Recognition for Detailed Prompt Construction

The Natural Language API can extract entities (e.g., names, dates, medical terms) from text, enabling prompt engineers to build prompts that incorporate detailed context.

Use Case: For a legal document analysis tool, the API can extract entities like *"contract date," "parties involved,"* and *"terms of agreement."* Prompt engineers can use these entities to craft prompts like:

"Highlight any changes to the terms of agreement for contracts signed in 2023."

This approach ensures that the prompt is highly specific, guiding the AI to focus on relevant content.

Sentiment Analysis for Tone-Adjusted Prompts

The Natural Language API's sentiment analysis feature helps engineers evaluate and adjust the tone of prompts to match user expectations.

Use Case: In a mental health support application, sentiment analysis can ensure that prompts are worded sensitively. For example:

Initial Prompt: *"Describe how you're feeling today."*

Sentiment Analysis Output: If the user's sentiment is negative, the follow-up prompt can be adjusted to: *"I'm here to help. Would you like to share more about what's been troubling you?"*

By incorporating sentiment analysis, engineers can design prompts that respond empathetically, improving user experience.

8.4. AutoML and Custom Model Training for Domain-Specific Prompt Engineering

AutoML enables developers to build custom models for text classification, sentiment analysis, and entity recognition, without needing deep expertise in machine learning. This capability is invaluable for creating domain-specific models that can interpret and respond to highly specialized prompts.

AutoML Text Classification for Optimizing Prompt Outputs

AutoML can be used to build custom text classification models that categorize user inputs and trigger specific prompt templates based on the classification.

Use Case: In a legal advisory application, a custom AutoML model can classify user queries into categories like *"contract review," "intellectual property,"* and *"employment law."* Based on the classification, the prompt might change to:

"For a contract review, please provide the key clauses you want to focus on."

This classification-based approach ensures that the prompts are contextually relevant and tailored to the user's needs.

8.5. Cloud Translation API for Multi-Language Prompt Engineering

Cloud Translation API enables real-time translation of prompts, expanding the reach of generative models across different languages and regions. Prompt engineers can use it to design multilingual AI applications, ensuring consistency and accuracy in prompt interactions across diverse audiences.

Translating Prompts for Multilingual Support

Engineers can use the Translation API to localize prompts, making it easier to deploy generative AI models in multilingual environments.

Use Case: A customer support chatbot can dynamically translate prompts like:

"Please describe your issue in detail," into multiple languages based on user preference.

This ensures that users receive prompts in their native language, improving comprehension and engagement.

8.6. BigQuery for Analyzing Prompt Performance and Optimization

BigQuery, Google Cloud's data warehouse, can be used to analyze prompt performance at scale. Prompt engineers can log prompt-response pairs, user interactions, and performance metrics in BigQuery, enabling deep analysis and optimization.

Using BigQuery for Prompt Evaluation and Refinement

Engineers can query historical data to identify patterns in how different prompts perform, revealing insights into which prompts generate the best user engagement and which need refinement.

Use Case: For a virtual assistant, engineers might analyze how users respond to different types of prompts in order to optimize greeting messages, escalation triggers, or question formats.

Feature Highlight: BigQuery's integration with tools like Data Studio allows engineers to create dashboards that visualize prompt performance, making it easy to track KPIs and iterate on prompt designs.

9. Azure Tools for Prompt Engineering

Azure provides a comprehensive set of tools to facilitate prompt engineering, helping developers refine inputs, optimize model responses, and analyze outcomes effectively. This section discusses Azure tools and services, including Azure OpenAI, Azure Machine Learning, Azure Cognitive Services, and other Azure resources that support prompt engineering workflows for Generative AI..

9.1. Azure OpenAI Service for Prompt Engineering

Azure OpenAI Service provides access to powerful models like GPT, Codex, and DALL-E, which can generate text, code, and images. Azure OpenAI Service integrates directly with Azure Machine Learning, enabling prompt engineering experimentation within a managed environment.

Setting Up Azure OpenAI Service

To begin using Azure OpenAI Service, provision the service in your Azure portal and create an API key for secure access.

```
# Azure CLI command to create an OpenAI resource
az cognitiveservices account create \
  --name "<YourOpenAIServiceName>" \
  --resource-group "<ResourceGroupName>" \
  --kind "OpenAI" \
  --sku "S0" \
  --location "<Location>"
```

Using OpenAI Models for Prompt Engineering

After configuring your service, you can interact with the model using the REST API or the Azure SDK.

```
# Python code snippet for generating responses using Azure OpenAI
import openai
import os

openai.api_key = os.getenv("OPENAI_API_KEY")
response = openai.Completion.create(
  model="text-davinci-003",
  prompt="Explain quantum computing in simple terms.",
  temperature=0.5,
  max_tokens=100
)
print(response.choices[0].text.strip())
```

Prompt Optimization with Azure OpenAI

Azure OpenAI Service offers features like temperature and top_p to control response creativity and diversity, aiding in prompt optimization.

Temperature: Controls randomness. A higher value makes responses more diverse.

Top_p: Limits response choices to a top probability subset.

Use Case: Optimizing for Customer Support Responses

Prompt engineering in customer support applications can be optimized by adjusting temperature and prompt phrasing to make responses polite, informative, and aligned with brand voice.

9.2. Azure Machine Learning for Experimentation and Fine-Tuning

Azure Machine Learning provides an environment for testing, fine-tuning, and managing Generative AI models, making it invaluable for iterative prompt engineering.

Setting Up Experimentation in Azure Machine Learning

Create an experiment in Azure Machine Learning to test multiple prompts and configurations in parallel.

from azureml.core import Workspace, Experiment

```
# Connect to your workspace
ws = Workspace.from_config()
experiment = Experiment(workspace=ws, name="prompt-engineering-
experiment")
# Define prompts and hyperparameters for testing
prompts = ["What is the weather today?", "Tell me about climate change."]
parameters = {"temperature": [0.5, 0.7], "max_tokens": [50, 100]}
```

Fine-Tuning Models with Custom Data

Azure Machine Learning enables fine-tuning OpenAI models using custom datasets, adapting models to specific domain languages and improving response accuracy.

```
# Example setup for fine-tuning with a custom dataset
from azureml.train.hyperdrive import HyperDriveConfig, choice
# Hyperparameter search for prompt testing
hyperdrive_config = HyperDriveConfig(
    run_config=run_config,
    hyperparameter_sampling=parameters,
    primary_metric_name="accuracy",
    primary_metric_goal="MAXIMIZE"
)
experiment = Experiment(workspace=ws, name="prompt-optimization")
experiment.submit(hyperdrive_config)
```

Use Case: Fine-Tuning a Model for Financial Data Interpretation

A financial institution could fine-tune an OpenAI model with industry-specific terms and responses to enhance its effectiveness in financial customer support and advisory.

9.3. Azure Cognitive Services for Text Analysis

Azure Cognitive Services, such as Text Analytics, can analyze AI-generated content to identify sentiment, key phrases, language, and entities. This helps evaluate prompt effectiveness by assessing whether responses align with desired outcomes.

Sentiment Analysis

Sentiment analysis allows you to measure the tone of generated responses, which is essential for applications like customer service.

```
from azure.ai.textanalytics import TextAnalyticsClient
from azure.core.credentials import AzureKeyCredential

# Initialize Text Analytics Client
credential = AzureKeyCredential("<YOUR_KEY>")
endpoint = "https://<your-text-analytics-resource>.cognitiveservices.azure.com/"
text_analytics_client = TextAnalyticsClient(endpoint=endpoint, credential=credential)

# Analyze sentiment
response = text_analytics_client.analyze_sentiment(["The service was excellent."])
print(response[0].sentiment)
```

Key Phrase Extraction

Extracting key phrases helps evaluate whether generated responses contain important information.

Use Case: Ensuring Consistency in Knowledge Responses

For a knowledge management application, Azure Cognitive Services can ensure AI-generated responses are informative and consistent by extracting and analyzing key terms and phrases.

9.4. Azure Synapse Analytics for Data Preparation

High-quality prompts require well-prepared data. Azure Synapse Analytics enables data preparation, transformation, and analysis, supporting efficient data processing and feeding relevant information into prompt engineering workflows.

Data Transformation with Azure Synapse

Azure Synapse allows ETL processing, transforming raw data into structured formats that can support prompt engineering. This ensures data used in Generative AI is relevant, clean, and well-structured.

```
-- SQL query in Azure Synapse to transform customer data
SELECT
    CONCAT(first_name, ' ', last_name) AS full_name,
    LOWER(email) AS email_address,
    CASE
        WHEN region IS NULL THEN 'Unknown'
        ELSE region
    END AS region
FROM customer_data
WHERE is_active = 1
```

Use Case: Preparing Customer Data for Personalized Prompts

For customer personalization, Azure Synapse can aggregate and transform user data to craft prompts tailored to individual preferences or past interactions.

9.5. Azure Data Factory for Automated Prompt Testing Pipelines

Azure Data Factory (ADF) facilitates data orchestration and automation, supporting batch processing of prompt engineering tests across various configurations.

Orchestrating Prompt Testing Pipelines

Automate prompt testing workflows by using Azure Data Factory to trigger model responses, capture outputs, and evaluate results.

```
// Azure Data Factory Pipeline JSON for automated prompt testing
{
  "name": "PromptTestingPipeline",
  "properties": {
    "activities": [
      {
        "name": "TestPrompt",
        "type": "ExecuteFunction",
        "functionAppName": "<FunctionAppName>",
        "functionName": "generateAIResponse",
        "inputs": [
          {
            "name": "prompt",
            "value": "What is the weather today?"
          },
          {
            "name": "temperature",
            "value": "0.7"
          }
        ]
      },
      {
        "name": "AnalyzeResponse",
        "type": "AzureMLBatchScore",
        "linkedServiceName": "<AMLWorkspaceLinkedService>"
      }
    ]
  }
}
```

Use Case: Large-Scale Prompt Testing for E-commerce

In an e-commerce setting, ADF pipelines can test multiple product recommendation prompts, capture responses, and analyze them to identify the best-performing prompts.

9.6. Azure Monitor for Prompt Performance Tracking and Alerts

Azure Monitor helps track the performance and accuracy of prompts, with metrics for latency, error rates, and model response times.

Setting Up Metrics and Alerts

Set up Azure Monitor to track API calls to the model and alert when performance thresholds are exceeded.

```
# PowerShell command to create an alert rule for monitoring API latency

New-AzMetricAlertRule -Name "PromptLatencyAlert" -ResourceGroupName "<ResourceGroup>" -
TargetResourceId "/subscriptions/<subscription-id>/resourceGroups/<resource-
group>/providers/Microsoft.CognitiveServices/accounts/<account-name>" -MetricName "Latency" -
Operator GreaterThan -Threshold 200 -WindowSize "PT1M"
```

Use Case: Real-Time Monitoring for Customer Support Chatbots

In customer support, Azure Monitor tracks model response times, alerting administrators to slowdowns that could affect user experience.

9.7. Azure Policy for Compliance in Prompt Engineering

Azure Policy enforces compliance across Generative AI resources, ensuring prompt engineering aligns with organizational standards and regulations.

Enforcing Data Compliance Policies

Ensure compliance by enforcing data usage and handling standards through Azure Policy, e.g., mandating data encryption and access controls for AI training datasets.

```
// Azure Policy Definition to enforce encryption in AI datasets
{
 "properties": {
  "displayName": "Require Encryption for AI Training Data",
  "policyType": "BuiltIn",
  "mode": "All",
  "parameters": {},
  "policyRule": {
   "if": {
    "allOf": [
```

```
{
    "field": "type",
        "equals": "Microsoft.Storage/storageAccounts"
      },
      {
        "field": "Microsoft.Storage/storageAccounts/encryption.keySource",
        "notEquals": "Microsoft.Keyvault"
      }
     ]
    },
    "then": {
      "effect": "deny"
    }
  }
 }
}
```

Use Case: Ensuring GDPR Compliance for User Data

A company handling EU user data can enforce Azure Policy to mandate encryption and data handling practices, ensuring GDPR compliance for prompt engineering processes.

10. Use-Cases

Application of Prompt Engineering in Customer Service Scenarios

Consider a customer service scenario where a company wants to deploy an AI-powered chatbot to handle customer inquiries, complaints, and support requests. Here's how prompt engineering can be integrated into the architecture:

Identifying Key Use Cases: The first step is identifying the common customer service scenarios, such as order tracking, product information, complaints, and returns.

Creating Prompt Templates:

"I want to track my order." → Response: "Please provide your order number, and I will check the status for you."

"I have a complaint about my recent purchase." → Response: "I'm sorry to hear that. Can you describe the issue in more detail? I'll do my best to help."

Prompt Chaining for Multi-Step Resolution:

Step 1: "Identify the nature of the issue." (e.g., delayed order, damaged product)

Step 2: "Provide a tailored solution based on the issue type." (e.g., "I see your order is delayed. Would you like a refund or to wait for delivery?")

Step 3: "Draft a closure message summarizing the solution and confirming next steps."

By integrating prompt engineering into the customer service workflow, companies can ensure that their AI models handle inquiries with the appropriate tone, accuracy, and efficiency, providing a seamless and satisfying customer experience.

Prompt Engineering for Healthcare Sector

Prompt engineering in the healthcare sector is used to enhance the interaction between AI systems and patients, medical professionals, or administrative staff by ensuring that the AI models produce contextually accurate, ethically sound, and reliable outputs. This use case focuses on how prompt engineering can be leveraged to design and optimize a *Virtual Healthcare Assistant (VHA)* for supporting clinical decision-making, automating administrative tasks, and engaging with patients.

Scenario: *Virtual Healthcare Assistant for Clinical and Patient Support*

Imagine a healthcare organization deploying a Virtual Healthcare Assistant (VHA) powered by a large language model like GPT-4 to assist doctors and nurses, facilitate patient interactions, and streamline clinical documentation. The VHA is expected to handle a variety of tasks, including patient triage, providing medical information, summarizing patient histories, and scheduling follow-up appointments. Effective prompt engineering is crucial in ensuring that the VHA communicates accurately, ethically, and in a manner that respects patient confidentiality.

1. Use Case Overview

The Virtual Healthcare Assistant (VHA) is integrated into a healthcare provider's system and performs three primary functions:

Patient Interaction and Triage: Engages with patients to collect initial information, understand symptoms, and recommend next steps based on the severity of the reported conditions.

Clinical Support for Healthcare Professionals: Assists doctors by summarizing patient histories, generating clinical notes, and suggesting possible diagnoses based on patient symptoms.

Administrative Support: Schedules appointments, sends reminders, and updates patient records.

Each function relies on carefully engineered prompts to ensure accurate outputs, compliance with healthcare regulations, and an optimal patient experience.

2. Patient Interaction and Triage

The VHA is designed to handle patient inquiries, collect symptoms, and provide preliminary recommendations. Prompt engineering is used to guide the model in asking the right questions, interpreting patient responses, and determining when to escalate to a human professional.

2.1. Symptom Collection and Analysis

The first interaction involves the VHA asking the patient about their symptoms and collecting relevant information. The prompts must be clear, empathetic, and non-intrusive to encourage patients to share accurate details.

Initial Symptom Prompt:

"Hello, I'm your virtual health assistant. Could you describe any symptoms you're currently experiencing? Please include details like how long they've persisted, their intensity, and any changes you've noticed."

This prompt sets a positive, open tone, encouraging patients to share comprehensive information.

2.2. Contextual Follow-Up Prompts

Depending on the initial response, the VHA needs to ask follow-up questions to gather more details and narrow down possible conditions.

Follow-Up Prompt Example:

"Thank you for sharing. When you say you have a headache, do you also experience any of the following: sensitivity to light, nausea, or dizziness?"

This prompt uses a list-based format, making it easy for the patient to provide additional information without feeling overwhelmed.

2.3. Risk Assessment and Next Steps

Once the VHA collects the symptom details, it must evaluate the severity and decide on the next steps. If the reported symptoms indicate a potentially serious condition, the VHA must escalate the case to a healthcare professional.

Prompt for Escalation:

"Based on the symptoms you've described, it's important to speak with a healthcare professional as soon as possible. Would you like me to connect you with a specialist or schedule an appointment for today?"

This ensures that the VHA takes appropriate action based on risk, maintaining patient safety and compliance with medical standards.

Best Practices for Symptom Triage Prompts:

Use empathetic language to establish trust and encourage honest communication.

Avoid making definitive diagnoses; instead, focus on suggesting potential conditions.

Design prompts to escalate appropriately based on symptoms that indicate serious health risks (e.g., chest pain, difficulty breathing).

3. Clinical Support for Healthcare Professionals

Prompt engineering helps the VHA function as a reliable assistant to doctors and nurses, summarizing patient information, generating clinical documentation, and suggesting potential diagnoses.

3.1. Summarizing Patient Histories

Doctors often need quick summaries of patient histories to make informed decisions. The VHA should be able to condense large volumes of patient data into concise, relevant summaries.

Prompt for Patient History Summarization:

"Generate a summary of the patient's history based on the following records. Focus on chronic conditions, recent hospital visits, and ongoing medications."

This prompt directs the VHA to focus on critical information, ensuring that doctors receive a clear, concise overview of the patient's health status.

3.2. Suggesting Potential Diagnoses

Based on patient-reported symptoms and history, the VHA can suggest possible diagnoses while ensuring that the suggestions are framed cautiously to avoid overconfidence in the AI's recommendations.

Prompt for Diagnosis Suggestion:

"Based on the patient's symptoms of fatigue, weight loss, and frequent urination, suggest possible conditions that may be relevant. Consider factors like patient age, medical history, and family history of diabetes."

The output should include a list of possible conditions, each accompanied by a brief rationale, and a disclaimer emphasizing the need for a professional medical evaluation.

Best Practices for Clinical Support Prompts:

Include clear constraints to ensure the AI does not make definitive medical claims.

Use structured prompts that highlight key data points (e.g., patient age, history) to guide the AI's suggestions.

Always include disclaimers, such as *"This is not a diagnosis; consult a healthcare provider for professional advice."*

4. Administrative Support for Healthcare Providers

The VHA can also automate routine administrative tasks, such as appointment scheduling, sending reminders, and updating patient records. Prompt engineering is used to ensure that these interactions are handled efficiently and with minimal errors.

4.1. Scheduling Appointments

Prompt Example:

"I can help you schedule an appointment. Are you looking for a general consultation, or would you like to see a specialist? Please specify your preferred date and time."

This prompt is direct and clear, guiding the patient through the scheduling process step-by-step.

4.2. Rescheduling or Canceling Appointments

Prompt Example:

"I see you have an appointment scheduled for [Date/Time]. Would you like to reschedule or cancel it? Please provide a new date and time if you're rescheduling."

After confirming the patient's choice, the VHA should respond with a confirmation message:

"Your appointment has been successfully rescheduled to [New Date/Time]. Would you like a reminder sent to your phone?"

4.3. Updating Patient Records

The VHA can help update patient records based on new information collected during virtual consultations or follow-up interactions.

Prompt for Updating Records:

"Based on the recent interaction, update the patient's record to include the new symptom information: 'Persistent cough for two weeks, fatigue, and mild fever.' Label this entry as 'Follow-up Visit Notes.'"

This prompt ensures that the AI follows the correct format and updates the EHR accurately.

Best Practices for Administrative Support Prompts:

Use clear, concise language to avoid confusion and ensure accurate scheduling.

Include prompts for confirmation to minimize errors (e.g., "Are you sure you want to cancel this appointment?").

Ensure that all interactions are logged and auditable for compliance purposes.

5. Security, Compliance, and Ethical Considerations

Prompt engineering in the healthcare sector must also address security and compliance, particularly around patient data privacy and adherence to regulations like HIPAA (Health Insurance Portability and Accountability Act).

5.1. Data Privacy and Security Prompts

Prompt Example for Sensitive Data Detection:

"If the input includes personally identifiable information (PII) such as Social Security numbers, phone numbers, or addresses, do not process the data and return an error message: 'Sensitive data detected. Please redact personal information before proceeding.'"

This ensures that the VHA adheres to privacy regulations and prevents sensitive information from being processed inadvertently.

5.2. Ensuring Ethical Use of AI in Healthcare

Prompts should be designed to maintain transparency and ethical behavior, such as avoiding biased language or ensuring that AI suggestions do not unduly influence clinical decisions.

Prompt for Ethical Interaction:

"If the patient asks for a specific treatment option that is not medically recommended, respond with: 'I'm unable to provide specific treatment recommendations. Please consult with your healthcare provider for personalized advice.'"

Best Practices for Security and Compliance Prompts:

Always include disclaimers when the AI provides health-related information.

Use prompt constraints to prevent the processing of sensitive patient data.

Ensure that prompts include clear guidance for escalation to human professionals when needed.

Conclusion: Prompt Engineering for a Safe and Effective Healthcare AI

Prompt engineering is a foundational capability for deploying AI systems in healthcare, ensuring that interactions are accurate, ethical, and aligned with medical standards. By carefully designing prompts for patient engagement, clinical support, and administrative tasks, healthcare organizations can deploy virtual assistants that enhance patient care, reduce clinician workload, and maintain compliance with healthcare regulations.

Generative AI Use Case for Legal Organizations

Generative AI is transforming industries by automating processes, enhancing decision-making, and generating content at scale. Legal organizations, traditionally bound by manual, time-intensive tasks, are increasingly benefiting from Generative AI (GenAI) applications. In this section, we will explore a specific use case of GenAI in legal organizations, focusing on contract generation, legal research, and document review. These tasks, which often consume a significant portion of legal professionals' time, can be streamlined using GenAI, allowing legal teams to work more efficiently and cost-effectively.

Scenario: Contract Generation and Legal Document Automation

In legal organizations, the generation, review, and management of contracts and other legal documents are core activities. These tasks, while critical, are often repetitive and require a high level of precision. Traditionally, legal teams draft contracts manually, relying on templates and customizing them based on client needs and specific case contexts. Generative AI can significantly enhance this process by automating the creation of contracts, ensuring compliance, reducing errors, and freeing up legal professionals to focus on higher-value work such as strategy and negotiation.

1. Use Case Overview

The application of Generative AI in this use case revolves around:

Contract Generation: Automatically generating customized contracts based on predefined templates, using client-specific information and contextual inputs.

Document Review: Automating the review of legal documents, including detecting inconsistencies, identifying risks, and suggesting improvements.

Legal Research and Insights: Accelerating legal research by using AI to analyze case law, statutes, and regulatory frameworks, offering contextual recommendations and summaries.

By integrating Generative AI into these workflows, legal organizations can speed up processes, reduce costs, and ensure that legal documents are accurate, consistent, and compliant with relevant regulations.

2. Contract Generation with Generative AI

2.1. Automated Contract Drafting

Generative AI can take on the task of drafting contracts by leveraging templates and predefined legal clauses. Legal teams can input specific details, and the AI can generate the full contract by filling in necessary sections, ensuring consistency across all legal documents.

Example: A legal firm needs to draft a *Non-Disclosure Agreement (NDA)* for a new client. Instead of manually inputting information, the AI can use a prompt like:

"Generate an NDA between [Client Name] and [Partner Name], covering intellectual property, data sharing, and a non-compete clause. Ensure the contract is compliant with [jurisdiction's] law."

The AI will then generate a draft of the NDA, populating key sections with information provided by the lawyer and ensuring that legal language remains precise and accurate.

2.2. Customization and Scalability

One of the major advantages of using GenAI in contract generation is its ability to customize contracts at scale. Legal organizations that deal with large volumes of contracts, such as real estate or corporate law firms, can automate the customization of contracts based on different client needs and jurisdictions.

Use Case: A corporate legal team responsible for managing mergers and acquisitions may need to generate numerous agreements in a short period. The AI can generate custom clauses for each transaction, ensuring that each contract is tailored to the specific merger while maintaining compliance with regulatory requirements.

Prompt Example:

"Create a shareholder agreement for a merger between Company A and Company B. The agreement should include provisions for minority shareholder protection, dividend distribution, and dispute resolution, based on U.S. corporate law."

The AI's ability to adapt templates and generate legally sound documents ensures that contract generation is scalable without sacrificing quality.

2.3. Clause Library and Compliance Checks

Legal teams often use clause libraries to ensure that contracts are consistent and compliant with current regulations. Generative AI can access these libraries to automatically include the appropriate clauses in a contract, based on the type of transaction and jurisdiction.

Use Case: For law firms that operate across multiple regions, compliance with local laws is crucial. A GenAI system can be integrated with a database of jurisdiction-specific clauses, ensuring that every contract is compliant with relevant local regulations.

Feature Highlight: By using a prompt like, *"Insert a force majeure clause compliant with New York state law,"* the AI can automatically incorporate the correct legal language without needing manual intervention, reducing the risk of compliance errors.

3. Document Review and Risk Assessment

In addition to generating contracts, Generative AI can assist in reviewing legal documents to detect potential risks, identify inconsistencies, and suggest improvements. The AI can analyze large volumes of text and flag problematic sections based on predefined risk parameters.

3.1. Reviewing and Comparing Contracts

Generative AI can review contracts in bulk and highlight discrepancies between different versions of contracts or compare them with standard templates. This capability is especially valuable in industries like real estate, where contracts often need to be compared against state-specific regulations or precedents.

Use Case: A legal team may need to review multiple leases for a commercial property acquisition. The AI can analyze the contracts and flag differences in lease duration, rent escalation clauses, or termination terms, providing a summary of key variations between contracts.

Prompt Example:

"Compare the attached commercial leases and highlight any discrepancies in rental terms, renewal conditions, and termination clauses."

By automating this comparison process, legal professionals can quickly identify issues that require further attention, streamlining the review process and minimizing manual effort.

3.2. Risk Detection and Mitigation

Generative AI can also analyze contracts for potential risks, such as ambiguous clauses, missing terms, or non-compliance with regulatory requirements. By integrating risk detection algorithms, AI can flag problematic areas and suggest modifications.

Use Case: During the review of a vendor agreement, the AI detects that the indemnification clause is overly broad, exposing the company to significant legal liability. The AI can flag the issue and suggest revisions, such as narrowing the scope of indemnification or specifying limits on liability.

Prompt Example:

"Review the indemnification clause in this vendor agreement and identify any risks related to liability or enforceability under U.K. contract law."

By automatically identifying risks and providing suggestions for improvement, AI ensures that legal teams can address issues proactively, reducing the chance of disputes or litigation.

4. Legal Research and Case Analysis

Legal research is a time-consuming but crucial part of legal practice. Generative AI can enhance legal research by quickly scanning vast databases of case law, regulations, and legal texts, summarizing key points, and offering relevant case precedents.

4.1. Accelerating Case Law Research

Generative AI can dramatically speed up the process of case law research by parsing through legal databases and presenting the most relevant precedents and rulings for a given case.

Use Case: A lawyer working on a trademark infringement case needs to find relevant precedents in U.S. intellectual property law. The AI can be prompted to search for and summarize key cases, such as:

"Find and summarize case law on trademark infringement in the tech industry, focusing on cases where the plaintiff was awarded damages."

The AI would provide summaries of the most relevant cases, along with citations and a brief analysis of the outcomes.

4.2. Regulatory Analysis and Compliance

Legal teams often need to ensure that contracts, business practices, and internal policies comply with ever-changing regulations. Generative AI can be used to monitor changes in regulatory frameworks and help legal professionals assess compliance.

Use Case: A financial services firm needs to ensure that its lending agreements comply with the latest federal regulations. The AI can scan new regulations and compare them against the company's existing agreements, flagging any areas of non-compliance.

Prompt Example:

"Analyze the latest changes in U.S. lending regulations and identify any sections of our standard loan agreement that need to be updated to ensure compliance."

By using AI for regulatory analysis, legal teams can stay up to date with the latest laws and ensure that their agreements are always compliant.

5. Benefits of Generative AI for Legal Organizations

5.1. Efficiency and Scalability

Generative AI significantly reduces the time required to draft, review, and customize legal documents. This scalability allows legal organizations to handle larger volumes of work without increasing headcount, thereby reducing costs and improving efficiency.

5.2. Improved Accuracy and Consistency

By automating document generation and review processes, AI minimizes the risk of human error, ensuring that contracts and legal documents are accurate, compliant, and consistent across the organization.

5.3. Enhanced Legal Research

AI-powered research tools can rapidly scan legal databases, reducing the time needed for case law research and allowing lawyers to focus on strategy rather than manual information gathering.

5.4. Risk Mitigation

AI's ability to detect and flag risky clauses or non-compliant sections of contracts ensures that legal teams can address issues before they become problems, reducing the risk of costly litigation.

Conclusion: Generative AI as a Game Changer for Legal Organizations

Generative AI is revolutionizing the legal industry by automating time-consuming tasks, enhancing accuracy, and improving efficiency. From drafting contracts to reviewing documents and conducting legal research, AI-driven tools allow legal professionals to focus on higher-value work, such as strategy, client relationships, and negotiations. By leveraging GenAI, legal organizations can streamline operations, reduce costs, and offer enhanced services to clients, positioning themselves at the forefront of innovation in the legal sector.

Conclusion

Prompt engineering is a pivotal capability for utilizing large language models effectively. As AI technology continues to develop, mastering this skill will enable individuals and organizations to unlock new levels of performance, creativity, and problem-solving potential

Cloud providers provides a robust ecosystem of tools and services that support every stage of prompt engineering, from model development to real-time deployment and monitoring. By leveraging these tools—organizations can design, deploy, and refine AI models that deliver highly accurate, contextually relevant, and secure outputs in diverse use cases, from healthcare to finance and beyond

The future of prompt engineering will see advancements in *automated prompt optimization*, *multi-modal prompting*, and *integrated prompt-programming languages*. As AI models evolve, prompt engineering will continue to be a foundational skill for AI practitioners and non-experts alike.

Chapter 9

Evolving the Data Strategy for Generative AI Maturity

Chapter Overview

The chapter explores the evolution of data strategies required to support Generative AI maturity across four distinct stages: **Foundational Data Management, Advanced Data Integration, AI-Driven Data Architectures,** and **Autonomous and Self-Optimizing Data Systems**. Each level is characterized by increasingly sophisticated data capabilities, cloud-native tools, and architecture patterns that enable organizations to unlock the full potential of Generative AI.

For each stage, the chapter presents detailed strategies, implementation guidelines, and industry examples to illustrate how real-world organizations are transforming their data ecosystems. Use cases span diverse industries such as **Retail**, **Healthcare**, **Finance**, and **Media**, providing practical insights into how these strategies can be applied to solve complex business challenges.

1. Foundational Data Management

Overview

Foundational Data Management focuses on creating a robust, secure, and scalable data infrastructure that supports consistent ingestion, integration, and storage. This stage is about establishing a solid data foundation that ensures data quality, governance, and security for downstream AI and analytics applications. Organizations at this level aim to build automated data pipelines, enforce compliance, and implement data governance frameworks to maintain traceability and trustworthiness.

Key Strategies

1. **Centralized Data Ingestion and Integration:**
    - Build unified data pipelines using **AWS Glue**, **Azure Data Factory**, or **Google Cloud Dataflow** to automate the ingestion of data from diverse sources.
    - Implement schema management and validation techniques to maintain data consistency and accuracy.
2. **Data Governance and Compliance Frameworks:**
    - Use data governance tools like **Azure Purview**, **AWS Lake Formation**, and **Google Cloud Data Catalog** to enforce data lineage tracking, catalog data assets, and establish data ownership.
    - Implement RBAC (Role-Based Access Control) and data masking policies to protect sensitive data and meet regulatory requirements.
3. **Secure and Scalable Data Storage Solutions:**
    - Leverage cloud-native storage solutions such as **AWS S3**, **Azure Blob Storage**, and **Google Cloud Storage** with encryption (e.g., **AWS KMS**, **Azure Key Vault**, **Google Cloud KMS**).
    - Implement disaster recovery and backup strategies using multi-region storage configurations to ensure data availability and resilience.

Industry Examples

1. **Retail:**
    - A global retail chain integrates sales, inventory, and customer interaction data using **AWS Glue** and stores it in an **AWS S3** data lake. The data is then processed and transformed using **AWS Glue** ETL jobs, making it available for real-time analytics in **Amazon Redshift**.
    - **Implementation Blueprint**: Set up an **S3-based data lake**, use **AWS Glue** for ETL, and load the cleaned data into **Amazon Redshift** for analysis.
2. **Healthcare:**
    - A hospital network deploys **Azure Purview** to track the lineage and usage of sensitive patient data. Data masking is implemented to protect PHI (Protected Health Information) when shared across departments, ensuring compliance with HIPAA regulations.
    - **Implementation Blueprint**: Deploy **Azure Purview** for data governance, use **Azure Data Factory** for secure data movement, and implement column-level security using **Azure Synapse Analytics**.
3. **Finance:**
    - A financial services firm uses **Google Cloud Dataflow** for real-time ingestion of transaction logs, implementing schema validation and data consistency checks before storing the data in a central **Google BigQuery** warehouse for compliance reporting.
    - **Implementation Blueprint**: Use **Google Cloud Pub/Sub** for data ingestion, apply transformations in **Dataflow**, and store the data in **BigQuery** with IAM controls for secure access.

2. Advanced Data Integration and Hybrid Architectures

Overview

At this stage, organizations enable hybrid data architectures that seamlessly integrate data from on-premises, cloud, and real-time sources. The focus is on building advanced data pipelines that support complex, cross-environmental data flows. This stage is essential for organizations that need to maintain data sovereignty, compliance, or leverage a mix of legacy systems and modern cloud solutions. The ability to integrate real-time data streams and hybrid architectures is crucial for enabling use cases like real-time personalization, cross-cloud analytics, and dynamic supply chain management.

Key Strategies

1. **Hybrid Data Integration Across Cloud and On-Premises:**
    - Use solutions like **Azure Arc**, **AWS Outposts**, and **Google Anthos** to enable consistent data management across cloud and on-premises environments.
    - Implement data replication and synchronization using **AWS Database Migration Service** or **Google Cloud Data Transfer Service**.
2. **Real-Time Data Streaming and Event-Driven Architectures:**
    - Use **AWS Kinesis**, **Azure Stream Analytics**, and **Google Cloud Pub/Sub** for low-latency data ingestion and processing.
    - Implement serverless event-driven services like **AWS Lambda** or **Azure Functions** to handle real-time data streams.
3. **Data Lakehouse Architectures for Unified Analytics:**
    - Implement **Delta Lake** on **Databricks** or **Google BigQuery Omni** to enable unified analytics across data lakes and data warehouses.
    - Use **AWS Redshift Spectrum** and **Azure Synapse** to build integrated querying capabilities across structured and unstructured data.

Industry Examples

1. **Retail:**
    - An e-commerce giant uses **Google Cloud Composer** to build a real-time recommendation system that integrates on-premises CRM data with cloud-hosted clickstream data. The system processes user behavior data in real-time and updates personalized recommendations dynamically.
    - **Implementation Blueprint**: Use **Google Cloud Composer** for orchestration, **Cloud Pub/Sub** for data ingestion, and **BigQuery** for real-time analytics.
2. **Healthcare:**
    - A healthcare provider integrates on-premises patient records with cloud-hosted analytics platforms using **Azure Arc**. This hybrid architecture allows them to perform real-time analytics on patient data without violating data residency regulations.
    - **Implementation Blueprint**: Use **Azure Arc** for cross-cloud orchestration, synchronize data using **Azure Data Factory**, and enable analytics in **Azure Synapse**.

3. **Finance:**
    - A global bank builds a fraud detection system using **AWS Kinesis** and **AWS Lambda** to process real-time transaction data. When anomalies are detected, the system triggers alerts and performs automated checks.
    - **Implementation Blueprint**: Implement **AWS Kinesis Data Streams** for real-time ingestion, use **Lambda** for event processing, and configure **AWS SNS** for alerting.

3. AI-Driven Data Architectures

Overview

In this phase, organizations evolve their data strategies to support AI-specific architectures optimized for machine learning workloads. The focus is on automating feature engineering, implementing data quality monitoring, and building AI-driven data pipelines that can handle large-scale training and inference. These architectures are designed to enable faster experimentation, reduce time-to-market, and support advanced AI use cases like predictive maintenance, fraud detection, and contextual recommendations.

Key Strategies

1. **Automated Feature Engineering and Data Transformation:**
    - Use tools like **AWS SageMaker Data Wrangler**, **Azure Machine Learning**, and **Google Vertex AI** to automate feature extraction and transformation.
    - Implement advanced feature techniques such as feature crossing and embedding-based transformations.
2. **Data Quality Monitoring and Drift Detection:**
    - Use AWS SageMaker Model Monitor, Azure Data Quality Services, and Google Vertex AI to track data quality and detect anomalies.
    - Implement drift detection algorithms like Kolmogorov-Smirnov tests to monitor distribution changes.
3. **AI-Specific Data Pipelines and Orchestration:**
    - Use orchestration tools like **Apache Airflow**, **AWS Step Functions**, or **Google Cloud Composer** to manage complex ML workflows.
    - Integrate with CI/CD tools like **Azure DevOps** or **AWS CodePipeline** to automate pipeline deployments.

Industry Examples

1. **Media:**
    - A streaming platform uses **AWS SageMaker Data Wrangler** to engineer features such as content engagement scores and viewer sentiment for building personalized content recommendations.
    - **Implementation Blueprint**: Use **AWS SageMaker Studio** for feature engineering, integrate with **SageMaker Feature Store**, and build ML models using **SageMaker Training Jobs**.

2. **Healthcare:**
   - A medical research organization builds an AI pipeline using **Google Vertex AI** to process genomic data and develop predictive models for disease progression.
   - **Implementation Blueprint**: Use **Vertex AI Pipelines** to automate feature engineering, data validation, and model training.

4. Autonomous and Self-Optimizing Data Systems

Overview

Autonomous and Self-Optimizing Data Systems represent the pinnacle of Generative AI maturity. These systems leverage real-time feedback loops, multi-agent learning, and advanced optimization strategies to adapt dynamically to changing conditions without human intervention. The goal is to build self-learning, context-aware systems that can continuously optimize data pipelines, resource allocation, and model performance, enabling use cases such as autonomous supply chain management, real-time trading systems, and adaptive healthcare solutions.

Key Strategies

1. **Self-Adapting Data Pipelines with Reinforcement Learning:**
   - Use reinforcement learning algorithms to dynamically adjust pipeline parameters based on real-time performance feedback.
   - Implement hierarchical feedback loops that use multi-level metrics to guide optimization.
2. **Context-Aware Processing and Dynamic Resource Allocation:**
   - Implement graph-based learning and contextual bandit algorithms for real-time context integration.
   - Use Kubernetes-based orchestration with **Kubeflow**, **Azure Kubernetes Service**, and **Google Kubernetes Engine** for dynamic scaling.
3. **Continuous Feedback Loops and Self-Learning Systems:**
   - Establish continuous feedback loops that monitor system performance, detect anomalies, and trigger automatic adjustments.
   - Use multi-agent systems to implement self-learning mechanisms that refine system behavior based on real-time interactions.

Industry Examples

1. **Retail:**
   - A global retail chain builds an autonomous supply chain management system using multi-agent RL to optimize inventory levels in real-time.
   - **Implementation Blueprint**: Implement a **multi-agent RL model** using **Azure Machine Learning**, set up a dynamic inventory management pipeline using **Azure Synapse**, and use **Azure Monitor** for real-time feedback.
2. **Finance:**
   - A trading platform uses self-adapting data pipelines to automatically optimize trading algorithms based on real-time market data.

- **Implementation Blueprint**: Use **Google Vertex AI** for real-time prediction, integrate with **BigQuery** for data storage, and build RL models using **Kubeflow Pipelines**.

Conclusion

By evolving their data strategies across these four stages, organizations can unlock the full potential of Generative AI. Each stage builds on the previous one, adding more capabilities and complexity, enabling companies to leverage advanced AI solutions for business transformation. Reaching the final stage of **Autonomous and Self-Optimizing Systems** allows businesses to create intelligent systems that continuously optimize and learn, driving substantial value and competitive advantage.

Chapter 10

Testing Strategy and Approach for Generative AI Implementations

Chapter Overview

Generative AI models, with their ability to create text, images, code, and even interactive applications, have opened new avenues for business innovation. However, testing these models is significantly more challenging than testing traditional software systems. The inherent variability in AI-generated outputs, the complexity of assessing subjective quality, and the increased risk of biases or security vulnerabilities make it essential to adopt a multi-layered and comprehensive testing strategy.

This chapter provides a detailed, step-by-step approach to building a robust testing strategy for Generative AI implementations, focusing on seven critical areas: **Functional Testing**, **Non-Functional Testing**, **Security Testing**, **Adversarial Testing**, **Performance Testing**, **Bias and Fairness Testing**, and **Developing a Robust Testing Framework**. Each section delves into best practices, methodologies, tools, and real-world industry use cases, providing a holistic framework for testing and validating Generative AI systems. By following the guidelines in this chapter, enterprises can ensure that their AI systems are reliable, secure, ethical, and aligned with business objectives.

1. Functional Testing for Generative AI Models

1.1. Overview

Functional testing focuses on verifying that the AI system performs its intended functions correctly and generates outputs that meet the defined requirements. In the context of Generative AI, this means validating the quality, relevance, and coherence of generated content (e.g., text, images, or code) under various input scenarios. Given the non-deterministic nature of Generative AI models, functional testing involves not just checking for correct outputs but also ensuring that the generated content is contextually appropriate and aligned with user expectations.

1.2. Key Functional Testing Strategies

1. **Input-Output Pair Testing**
    - Test predefined input-output pairs to validate that the model's output is relevant, accurate, and meets business requirements.
    - **Example**: A Generative AI chatbot is given a set of standard customer queries, and its responses are evaluated against a predefined set of acceptable responses for relevance, tone, and accuracy.
2. **Edge Case Testing**
    - Evaluate how the model handles rare or ambiguous inputs that fall outside typical scenarios. This helps identify how the model reacts to unexpected inputs.
    - **Example**: An AI image generation model is tested with prompts describing surreal or impossible scenarios (e.g., "a cat swimming underwater with a fish tail") to see if it produces a coherent image.
3. **Regression Testing**
    - Ensure that updates to the model, such as retraining or fine-tuning, do not negatively impact its performance on previously validated scenarios.
    - **Example**: A text summarization model is tested on a standard benchmark dataset after being retrained with new data to verify that its summarization quality has not degraded.

1.3. Tools and Frameworks for Functional Testing

- **AWS SageMaker Clarify**: To monitor model outputs for accuracy and compliance with predefined criteria.
- **Azure Machine Learning Testing Suite**: For automated testing and validation of model behavior.
- **Google Cloud AutoML**: For automated input-output validation and model performance testing.

1.4. Best Practices for Functional Testing

- Define a range of acceptable outputs rather than a single correct answer, given the variability in Generative AI outputs.
- Include subject matter experts in the testing process to provide qualitative assessments where necessary.
- Implement automated testing pipelines to continuously validate model behavior as new data is introduced.

2. Non-Functional Testing for Generative AI

2.1. Overview

Non-functional testing evaluates attributes of the Generative AI system that are not related to specific outputs but are essential for its overall performance, reliability, and user experience. This includes aspects such as performance, scalability, resource utilization, and operational efficiency. Non-functional testing is particularly important for production-grade Generative AI models that must operate at scale and respond to dynamic user demands.

2.2. Key Non-Functional Testing Strategies

1. **Performance Testing**
   - Measure the system's latency, throughput, and response times under different load conditions.
   - **Example**: A text generation API is tested to ensure it responds to user queries within 200 milliseconds, even when processing multiple concurrent requests.
2. **Scalability Testing**
   - Evaluate the system's ability to scale dynamically with increased or decreased demand.
   - **Example**: An AI-driven recommendation engine is tested by simulating varying levels of user activity to ensure it can handle traffic spikes without performance degradation.
3. **Resource Utilization Analysis**
   - Assess the model's consumption of CPU, GPU, and memory resources to optimize infrastructure costs.
   - **Example**: A deep learning model for image generation is analyzed for GPU usage during peak processing periods to identify potential cost-saving optimizations.

2.3. Tools and Frameworks for Non-Functional Testing

- **AWS CloudWatch**: For real-time monitoring of performance metrics and resource usage.
- **Azure Monitor**: For tracking operational metrics and system health.
- **Google Cloud Profiler**: For analyzing resource consumption and identifying performance bottlenecks.

2.4. Best Practices for Non-Functional Testing

- Implement real-time monitoring and alerts to detect performance degradation or resource overutilization.
- Use containerization and orchestration tools (e.g., Kubernetes) to manage resource allocation dynamically.
- Perform load testing regularly, especially before deploying models to production environments.

3. Security Testing for Generative AI Models

3.1. Overview

Security testing is essential to safeguard Generative AI models from adversarial attacks, data breaches, and unauthorized access. Given the sensitivity of data used in training and the potential for models to be manipulated, security testing must be a continuous process that spans model development, deployment, and maintenance.

3.2. Key Security Testing Strategies

1. **Penetration Testing**
    - Simulate real-world attacks to identify vulnerabilities in the model and its deployment environment.
    - **Example**: A penetration test is conducted on a voice recognition model to identify potential vulnerabilities in the API that could allow unauthorized access.
2. **Data Leakage Prevention**
    - Test for scenarios where sensitive information may be exposed by the model.
    - **Example**: A customer service AI is tested to ensure it does not inadvertently reveal personal data (e.g., names, account numbers) in its responses.
3. **Access Control Testing**
    - Ensure that only authorized users and applications can access sensitive models and data.
    - **Example**: An AI model used for fraud detection is tested to verify that only approved financial analysts can access its predictive outputs.

3.3. Tools and Frameworks for Security Testing

- **Microsoft Counterfit**: For conducting automated security assessments and adversarial testing.
- **AWS IAM and GuardDuty**: For monitoring access control and detecting unusual activities.
- **Google Cloud Security Command Center**: For managing and mitigating security risks.

3.4. Best Practices for Security Testing

- Implement multi-layered security controls, including encryption, identity management, and secure access protocols.
- Conduct regular security audits and update models and systems to address emerging threats.
- Use adversarial training techniques to enhance model robustness against security threats.

4. Adversarial Testing for Generative AI Implementations

4.1. Overview

Adversarial testing involves exposing the model to carefully crafted malicious inputs to evaluate its robustness against adversarial attacks. These attacks aim to manipulate model outputs, degrade performance, or extract sensitive information. Adversarial testing is critical for AI systems that operate in high-risk environments or handle sensitive data.

4.2. Key Adversarial Testing Strategies

1. **Evasion Testing**
    - Generate adversarial inputs to deceive the model into making incorrect predictions.
    - **Example**: Small perturbations are added to a benign image to cause a facial recognition model to misclassify the person.

2. **Data Poisoning Testing**
   - Inject malicious data into the training set to corrupt the model's outputs.
   - **Example**: A customer sentiment analysis model is poisoned with skewed data to make it favor specific products.

3. **Model Inversion Testing**
   - Attempt to reconstruct sensitive inputs by observing the model's outputs.
   - **Example**: An attacker queries a healthcare diagnostic model to extract private patient information.

4.3. Tools and Frameworks for Adversarial Testing

- **IBM Adversarial Robustness Toolbox**: For crafting and testing adversarial attacks.
- **Google's CleverHans Library**: For generating adversarial examples and evaluating model resilience.
- **Foolbox**: For adversarial testing across different deep learning frameworks.

4.4. Best Practices for Adversarial Testing

- Use a combination of white-box, black-box, and grey-box testing approaches to simulate various threat scenarios.
- Incorporate adversarial testing into the continuous integration/continuous deployment (CI/CD) pipeline.
- Regularly update models and retrain them with adversarial examples to improve robustness.

5. Performance Testing for Generative AI Models

5.1. Overview

Performance testing ensures that the AI system meets the expected levels of responsiveness, throughput, and resource efficiency. This type of testing is particularly important for real-time or interactive Generative AI applications, where delays or resource overuse can significantly impact user experience.

5.2. Key Performance Testing Strategies

1. **Response Time Testing**
   - Measure the time taken by the model to generate outputs for different types of inputs.
   - **Example**: An AI-powered virtual assistant is tested to ensure that it responds to user queries within an acceptable time frame.
2. **Throughput Testing**
   - Evaluate the number of requests the model can handle per second without compromising performance.
   - **Example**: A text generation API is stress-tested with thousands of concurrent requests to evaluate throughput.
3. **Latency Testing**
   - Assess the delay between input and output generation, especially for real-time applications.

- **Example**: A language translation model is tested for latency during live conversations to ensure seamless user experience.

5.3. Tools and Frameworks for Performance Testing

- **AWS CloudWatch**: For tracking latency, throughput, and response times.
- **Azure Application Insights**: For monitoring model performance and detecting bottlenecks.
- **Google Cloud Profiler**: For optimizing model performance and resource usage.

5.4. Best Practices for Performance Testing

- Conduct performance testing under various scenarios, including peak load and resource-constrained environments.
- Use profiling tools to identify performance bottlenecks and optimize model configurations.
- Implement real-time monitoring and alerts to detect and address performance issues proactively.

6. Bias and Fairness Testing for Generative AI Models

6.1. Overview

Bias and fairness testing ensures that Generative AI models produce outputs that are free from unfair biases related to gender, race, age, or other protected attributes. This type of testing is essential for maintaining the ethical integrity of AI systems and ensuring compliance with legal and regulatory standards.

6.2. Key Bias and Fairness Testing Strategies

1. **Fairness Metrics Analysis**
   - Measure fairness metrics such as equal opportunity, disparate impact, and demographic parity.
   - **Example**: A hiring recommendation model is tested to ensure that it provides equal opportunities for all demographic groups.
2. **Subgroup Performance Analysis**
   - Assess the model's performance across different demographic subgroups to identify potential disparities.
   - **Example**: A loan approval model is tested to ensure it does not disproportionately reject applicants from specific ethnic groups.
3. **Bias Mitigation Techniques**
   - Implement bias mitigation techniques and evaluate their effectiveness in reducing unfair biases.
   - **Example**: An AI-based hiring system is adjusted to minimize bias in resume screening, and the effectiveness of the mitigation is validated.

6.3. Tools and Frameworks for Bias and Fairness Testing

- **IBM AI Fairness 360**: For evaluating and mitigating bias in AI models.
- **Microsoft Fairlearn**: For assessing fairness metrics and implementing mitigation techniques.
- **Google's What-If Tool**: For visualizing and analyzing model fairness and subgroup performance.

6.4. Best Practices for Bias and Fairness Testing

- Involve diverse stakeholders in the testing process to provide comprehensive perspectives on fairness.
- Regularly audit models for fairness and update them to address emerging biases.
- Use a combination of quantitative and qualitative measures to assess fairness comprehensively.

7. Developing a Robust Testing Framework for Generative AI

7.1. Create a Testing Strategy Aligned with Business Goals

Define a clear testing strategy that aligns with business objectives and addresses specific risks and requirements of the application domain. Tailor the strategy based on whether the focus is on performance optimization, security testing, or ethical compliance.

7.2. Leverage Automation Tools for Continuous Testing and Monitoring

Use cloud-native automation tools to continuously test and monitor AI models throughout their lifecycle. Implement CI/CD pipelines to automate testing and reduce manual overhead.

7.3. Establish a Feedback Loop for Continuous Improvement

Incorporate testing results into a feedback loop to refine models, improve robustness, and address emerging risks. Engage stakeholders regularly to review testing outcomes and update testing strategies.

7.4. Implement Domain-Specific Test Scenarios

Design test scenarios that reflect the unique challenges and requirements of your industry, such as compliance testing in finance or ethical testing in healthcare.

7.5. Document Testing Results and Maintain Traceability

Maintain detailed documentation of testing procedures, results, and decisions to ensure traceability and accountability throughout the AI lifecycle.

Conclusion

A comprehensive testing strategy is essential for ensuring the reliability, security, and ethical alignment of Generative AI systems. By adopting the testing methodologies outlined in this chapter and leveraging cloud-native tools and frameworks, enterprises can deploy robust AI solutions that are safe, secure, and aligned with business and societal expectations.

Chapter 11

The Relevance of MLOps for Generative AI

Machine Learning Operations (MLOps) has become a crucial enabler for successfully deploying, monitoring, and scaling machine learning models in real-world environments. MLOps provides a structured approach to managing the complexities of machine learning workflows, integrating principles from DevOps, data engineering, and machine learning. While MLOps is relevant to all types of AI, it is especially significant for **Generative AI** models due to their size, complexity, and specific operational challenges.

Generative AI models—such as language models, generative adversarial networks (GANs), and transformer-based architectures—often have billions of parameters, require substantial computational resources, and need meticulous handling to ensure consistent performance. Implementing MLOps for these models is critical to streamline the end-to-end process, from data ingestion and model training to deployment, monitoring, and continuous optimization.

In this chapter, we will explore why MLOps is relevant for Generative AI, outline the key stages of the MLOps lifecycle, and provide an overview of the tools and services offered by AWS, Google Cloud, and Microsoft Azure for implementing MLOps in Generative AI projects.

1. Why MLOps is Critical for Generative AI

Generative AI models present unique challenges compared to traditional machine learning models. Their complexity, large scale, and varied use cases make them harder to deploy and maintain. MLOps frameworks address these challenges by automating the workflows, managing version control, and ensuring that models remain accurate and relevant over time.

1.1. Addressing the Complexity of Generative AI Models

Generative AI models, such as GPT-3, DALL-E, and PaLM, are built using advanced architectures like transformers, which are computationally intensive and require substantial resources for training. Managing these models effectively requires distributed training strategies, scalable storage, and robust deployment infrastructure.

- **Scalability**: MLOps frameworks provide the infrastructure to scale training and inference across multiple GPUs and TPUs.
- **Automation**: Automating the deployment and retraining of these models helps manage their complexity.

1.2. Enabling Experimentation and Continuous Improvement

Generative AI models need frequent experimentation to refine outputs, adjust hyperparameters, and optimize performance. MLOps supports this by tracking experiments, managing configurations, and providing tools for automated tuning.

- **Experiment Management**: Tools like Google Cloud's **Vertex AI Pipelines**, Amazon SageMaker's **Experiment Tracking**, and Azure's **ML Studio** help track and compare multiple experiments.
- **Hyperparameter Optimization**: Services like **Amazon SageMaker Automatic Model Tuning** and **Azure Automated Machine Learning** automate hyperparameter tuning to identify the best configurations.

1.3. Managing Version Control and Model Lineage

Tracking the lineage of Generative AI models is complex due to their iterative development process. MLOps tools help manage the versions of datasets, models, and code, ensuring reproducibility and compliance.

- **Model Versioning**: AWS provides **SageMaker Model Registry** for tracking model versions, while Google Cloud and Azure offer similar functionality through **Vertex AI Model Registry** and **Azure ML Model Management**.

1.4. Ensuring Robust Deployment and Monitoring

Generative AI models must be continuously monitored for performance, drift, and potential biases. MLOps frameworks enable automated monitoring and alerting, making it easier to identify and address issues before they impact production systems.

- **Real-Time Monitoring:** AWS SageMaker Model Monitor, Google Cloud's Vertex AI Monitoring, and Azure's ML Monitoring offer robust monitoring capabilities.

2. The MLOps Lifecycle for Generative AI

The MLOps lifecycle for Generative AI includes several key stages, each with unique requirements and best practices. Implementing a structured lifecycle ensures that Generative AI models are built, deployed, and maintained systematically.

2.1. Data Management and Preparation

Data is the foundation of any AI system. For Generative AI, data management is particularly challenging due to the large volumes and diverse formats involved (text, images, audio).

- **Data Preparation:** Use AWS Glue, Google Cloud Dataprep, or Azure Data Factory to clean, transform, and prepare data.
- **Data Storage:** Store large datasets in scalable storage solutions like Amazon S3, Google Cloud Storage, or Azure Blob Storage.
- **Data Versioning:** Implement data versioning using tools like DVC (Data Version Control) or built-in features in Google Cloud Vertex AI and Azure Machine Learning.

2.2. Experimentation and Model Building

Generative AI models require extensive experimentation to refine their outputs. MLOps tools provide experiment management and versioning capabilities to track hyperparameters, configurations, and model variations.

- **Experiment Tracking:** Use Amazon SageMaker Experiments, Google Cloud Vertex AI Experiments, or Azure ML Studio to manage experiments.
- **Notebook Environments:** Use managed environments like AWS SageMaker Studio, Google Cloud AI Notebooks, or Azure Notebooks for collaborative development.

2.3. Model Training and Tuning

Training Generative AI models is resource-intensive, requiring optimized infrastructure and distributed training strategies.

- **Distributed Training:** Use services like Amazon SageMaker Distributed Training, Google Cloud TPU Pods, or Azure Machine Learning ParallelRunStep to scale training across multiple nodes.
- **Hyperparameter Tuning:** Automate hyperparameter tuning using AWS SageMaker Tuning, Google Cloud Vertex AI Vizier, or Azure HyperDrive.

2.4. Model Deployment

Deploying Generative AI models involves setting up scalable and reliable endpoints for real-time inference or batch processing.

- **Managed Endpoints:** Use Amazon SageMaker Endpoints, Google Cloud AI Platform Prediction, or Azure Machine Learning Endpoints for deploying models.
- **Serverless Inference:** For lightweight models, consider serverless options like AWS Lambda, Google Cloud Functions, or Azure Functions.

2.5. Continuous Integration and Continuous Deployment (CI/CD)

CI/CD pipelines automate the deployment of Generative AI models, ensuring that every change to the code, data, or model triggers automated testing and deployment.

- **CI/CD Tools:** Use AWS CodePipeline, Google Cloud Build, or Azure Pipelines to automate CI/CD workflows.
- **Integration with Version Control:** Connect with GitHub or GitLab for automated version control and deployment.

2.6. Monitoring and Governance

Monitoring is crucial for detecting performance degradation, model drift, and biases in Generative AI models.

- **Model Monitoring:** Use Amazon SageMaker Model Monitor, Google Vertex AI Model Monitoring, or Azure Monitor for real-time monitoring.
- **Bias Detection:** Use SageMaker Clarify, Google's What-If Tool, or Azure Fairness Toolkit to detect and mitigate biases.

2.7. Continuous Learning and Retraining

As new data becomes available, retraining Generative AI models is necessary to keep them relevant and accurate.

- **Automated Retraining Pipelines:** Use AWS Step Functions, Google Cloud Vertex AI Pipelines, or Azure Machine Learning Pipelines for automated retraining.
- **Active Learning:** Implement active learning strategies to prioritize data for labeling and retraining.

3. AWS Tools for MLOps in Generative AI

AWS offers a suite of tools specifically designed to streamline MLOps for Generative AI:

- **Amazon SageMaker Studio**: An integrated development environment (IDE) that provides tools for model development, training, and deployment.
- **Amazon SageMaker Pipelines**: A fully managed CI/CD service for automating end-to-end ML workflows.
- **Amazon SageMaker Model Monitor**: Provides continuous monitoring for detecting data and model drift.
- **AWS Glue**: A data integration service for transforming and cleaning large datasets.
- **Amazon SageMaker Clarify**: Offers tools for bias detection and explainability.

4. Google Cloud Tools for MLOps in Generative AI

Google Cloud provides a comprehensive set of services for MLOps:

- **Vertex AI**: A unified platform for managing the entire ML lifecycle, including training, deployment, and monitoring.
- **Vertex AI Pipelines**: Orchestrates complex ML workflows with support for Kubeflow integration.
- **Vertex AI Model Monitoring**: Continuously monitors models for drift and anomalies.
- **BigQuery ML**: Allows training and deployment of models directly within BigQuery.
- **What-If Tool**: Helps in understanding and mitigating biases in models.

5. Microsoft Azure Tools for MLOps in Generative AI

Azure offers several robust tools for MLOps:

- **Azure Machine Learning**: An enterprise-grade platform for building, deploying, and managing machine learning models.
- **Azure ML Pipelines**: Automates end-to-end machine learning workflows.
- **Azure Model Monitoring**: Monitors model performance and detects drift.
- **Azure Fairness Toolkit**: Helps in evaluating and mitigating biases in models.
- **Azure Databricks**: Provides a collaborative platform for building scalable ML models.

6. Best Practices for Implementing MLOps in Generative AI

Adhering to best practices in MLOps is essential for ensuring the smooth and reliable operation of Generative AI models. Below are key best practices specifically tailored for implementing MLOps in Generative AI workflows:

6.1. Modular Pipeline Design

Modular pipeline design involves building machine learning workflows in distinct, reusable components that can be easily swapped or updated without affecting the entire pipeline. This design approach is particularly useful for Generative AI models, where frequent experimentation and adjustments are common.

- **Use Case**: Separating data preprocessing, model training, evaluation, and deployment into distinct steps enables easier debugging and optimization.
- Implementation:
  - Use tools like SageMaker Pipelines, Vertex AI Pipelines, or Azure Machine Learning Pipelines to define modular components.
  - Implement a component-based structure in tools like Kubeflow Pipelines, which offers flexibility for complex workflows.

6.2. Data and Model Versioning

Versioning is crucial for tracking changes and ensuring reproducibility in machine learning workflows. For Generative AI, where model behavior can be highly sensitive to minor data changes, versioning is particularly important.

- **Data Versioning:** Use tools like DVC (Data Version Control) or built-in capabilities of Google Vertex AI and Azure Machine Learning to track dataset versions.
- **Model Versioning:** Store multiple versions of models in registries like AWS SageMaker Model Registry, Google Vertex AI Model Registry, or Azure ML Model Management.
- **Use Case:** When fine-tuning Generative AI models, maintain clear version histories for each model iteration to track performance changes.

6.3. Automated Testing and Validation

Automating the testing and validation of Generative AI models helps catch issues early and ensures that the models perform as expected before deployment.

- **Unit Testing**: Create unit tests for key components like data preprocessing, feature engineering, and model training scripts.
- **Integration Testing**: Test the integration between different modules in the pipeline, such as feature engineering and model training.
- **End-to-End Testing**: Simulate real-world scenarios to validate the entire workflow.
- **Tooling**: Use **AWS CodeBuild**, **Google Cloud Build**, or **Azure DevOps** to integrate testing into your CI/CD pipelines.

6.4. Automation of Repetitive Tasks

Automation helps reduce manual intervention, improve efficiency, and minimize the risk of human errors in complex Generative AI workflows.

- **Use Automation Tools**: Implement tools like **AWS Step Functions**, **Google Cloud AI Pipelines**, or **Azure Machine Learning Pipelines** to automate repetitive tasks such as data ingestion, training, and deployment.
- **Hyperparameter Optimization**: Use services like **Amazon SageMaker Automatic Model Tuning**, **Google Vertex AI Vizier**, or **Azure HyperDrive** for automated hyperparameter tuning.

6.5. Robust Monitoring and Alerting

Monitoring is crucial for ensuring that Generative AI models remain accurate, reliable, and free from unintended biases. Implement a robust monitoring framework that covers the following areas:

- **Model Performance**: Track metrics like accuracy, perplexity, F1-score, and response times using tools like **Amazon CloudWatch**, **Google Cloud Monitoring**, or **Azure Monitor**.
- **Data Drift and Concept Drift**: Use **SageMaker Model Monitor**, **Google Vertex AI Monitoring**, or **Azure ML Model Monitoring** to detect shifts in input data or model behavior.
- **Bias and Fairness**: Continuously evaluate the model for fairness and bias using **SageMaker Clarify**, **Google's What-If Tool**, or **Azure Fairness Toolkit**.

6.6. Implementing Continuous Learning

Continuous learning involves periodically updating models as new data becomes available to keep them relevant and accurate.

- **Automated Retraining Pipelines:** Set up retraining pipelines using AWS Step Functions, Google Cloud Vertex AI Pipelines, or Azure Machine Learning Pipelines.
- **Active Learning:** Use active learning strategies to identify and prioritize data points that are most valuable for improving model performance.
- **Human-in-the-Loop:** Implement human-in-the-loop systems to validate model outputs and refine model behavior over time.

6.7. Secure Access and Compliance

Security and compliance are critical aspects of MLOps, especially when dealing with sensitive data in regulated industries like finance and healthcare.

- **Role-Based Access Control (RBAC)**: Implement RBAC using **AWS IAM**, **Google Cloud IAM**, or **Azure Active Directory** to control access to sensitive data and models.
- **Data Encryption**: Ensure data at rest and in transit is encrypted using tools like **AWS KMS (Key Management Service)**, **Google Cloud Key Management**, or **Azure Key Vault**.
- **Compliance Audits**: Use logging and monitoring tools to create audit trails and ensure compliance with regulations like GDPR, CCPA, or HIPAA.

7. Tools and Services from AWS, Google Cloud, and Microsoft Azure for MLOps in Generative AI

Each cloud provider offers a range of tools and services designed to support MLOps workflows specifically for Generative AI. Below, we explore the key offerings from AWS, Google Cloud, and Microsoft Azure:

7.1. AWS Tools for MLOps in Generative AI

- **Amazon SageMaker**: A comprehensive platform for building, training, and deploying machine learning models.
  - **SageMaker Studio**: An IDE for ML development, offering features like experiment tracking, debugging, and model monitoring.
  - **SageMaker Pipelines**: Provides CI/CD capabilities for building and automating ML workflows.
  - **SageMaker Clarify**: Detects bias and explains model predictions.
  - **SageMaker Model Monitor**: Continuously monitors models for drift and performance degradation.
  - **AWS Step Functions**: Automates workflows and orchestrates complex machine learning pipelines.

7.2. Google Cloud Tools for MLOps in Generative AI

- **Google Cloud Vertex AI**: A unified platform for managing the complete machine learning lifecycle.
  - **Vertex AI Pipelines**: Automates ML workflows, with integration for CI/CD.
  - **Vertex AI Model Monitoring**: Continuously monitors models for performance and drift.
  - **What-If Tool**: Analyzes models for fairness and bias.
  - **Kubeflow Pipelines**: An open-source tool for building and deploying portable, scalable ML workflows.
  - **BigQuery ML**: Enables training and deploying models directly in the BigQuery environment.

7.3. Microsoft Azure Tools for MLOps in Generative AI

- **Azure Machine Learning**: An enterprise-grade service for building, training, and deploying ML models.
  - **Azure ML Pipelines**: Automates ML workflows from data preparation to deployment.
  - **Azure ML Model Registry**: Manages model versions and tracks model lineage.

- **Azure Fairness Toolkit**: Evaluates and mitigates biases in models.
- **Azure Monitor**: Provides real-time monitoring and alerting for deployed models.
- **Azure DevOps**: Integrates with Azure ML to enable CI/CD pipelines for machine learning.

8. Conclusion

MLOps is essential for successfully managing the complex and resource-intensive workflows associated with Generative AI. By adopting best practices and leveraging the tools provided by cloud platforms like AWS, Google Cloud, and Microsoft Azure, organizations can build scalable, reliable, and compliant Generative AI systems. Implementing MLOps not only accelerates the deployment and iteration of AI models but also ensures that these models remain ethical, fair, and high-performing throughout their lifecycle.

As Generative AI continues to evolve, the need for robust MLOps frameworks will only become more critical. This chapter provides a comprehensive overview of the relevance, lifecycle, tools, and best practices for MLOps in Generative AI, equipping enterprises to navigate the challenges of deploying and maintaining sophisticated AI systems at scale..

Chapter 12

Role of Governance in Generative AI

Chapter Overview

Governance is critical to ensuring that Generative AI initiatives are developed and deployed responsibly within the enterprise. As AI systems become more integrated into business processes, the complexity of managing these initiatives increases. Effective governance frameworks provide a structured approach to navigating risks, ensuring compliance, and aligning AI projects with organizational goals.

The chapter breaks down the evolution of governance across four distinct stages of Generative AI maturity: **Initial Deployment**, **Intermediate Deployment**, **Advanced Deployment**, and **Mature AI Adoption**. Each stage is characterized by unique challenges and governance requirements, which are addressed through tailored strategies and best practices. Through detailed industry examples, this chapter illustrates how organizations can leverage governance to unlock the full potential of Generative AI while minimizing risks.

1. Understanding the Role of Governance in Generative AI Maturity

1.1. Why Governance is Critical for Generative AI Maturity

Governance serves as the backbone of any mature Generative AI system. As organizations implement AI to drive business outcomes, governance frameworks ensure that AI initiatives are compliant, ethical, and aligned with strategic objectives. With the growing adoption of Generative AI, governance is crucial for mitigating risks such as:

1. **Bias and Discrimination**: Unchecked models can reinforce or amplify biases present in training data.
2. **Lack of Transparency**: Without proper documentation and explainability, AI systems can become "black boxes," making it difficult to understand their decisions.

3. **Compliance Violations**: AI systems must adhere to regulatory requirements like GDPR, HIPAA, and CCPA, and governance frameworks help ensure this compliance.
4. **Operational Risks**: Governance mitigates risks like model drift, data leakage, and unintended use cases, ensuring AI remains reliable and aligned with its intended purpose.

1.2. Key Components of Generative AI Governance

1. **Data Governance**: Establishing rules and processes for managing data quality, security, privacy, and access control.
2. **Model Governance**: Frameworks for managing the lifecycle of AI models, including versioning, validation, and monitoring.
3. **Compliance Management**: Ensuring AI systems meet legal and regulatory requirements, such as data privacy laws and industry-specific regulations.
4. **Ethical Governance**: Addressing fairness, accountability, and transparency to ensure AI systems align with organizational values and societal expectations.
5. **Accountability and Transparency**: Defining roles and responsibilities for AI-related decisions and providing mechanisms for auditing and reporting.

2. Evolution of Governance Across the Generative AI Maturity Curve

2.1. Stage 1: Initial Deployment – Establishing Foundational Governance

Overview

During the initial stage of Generative AI adoption, organizations focus on small-scale pilots and proof-of-concept projects. Governance frameworks at this stage are nascent, with a primary focus on defining foundational policies and controls. The aim is to set the groundwork for responsible AI experimentation by establishing basic data governance policies, creating an ethics committee, and implementing preliminary compliance checks.

Governance Objectives at Stage 1

1. **Define Data Ownership and Access Control Policies**: Establish who owns the data, who can access it, and under what conditions.
2. **Implement Baseline Compliance and Security Standards**: Develop initial compliance guidelines to prevent data breaches and misuse.
3. **Create Ethical Review Processes**: Establish an ethics committee to evaluate the potential risks and societal impact of AI projects.

Key Strategies

- Data Governance Policies:
  - Use Azure Purview, AWS Glue Data Catalog, or Google Cloud Data Catalog to create a unified inventory of data assets.
  - Define data ownership and access control policies, ensuring sensitive data is handled responsibly.

- Basic Compliance Frameworks:
  - Implement compliance checks using **AWS IAM**, **Azure AD**, or **Google Cloud IAM** to restrict access to data and models based on user roles.
  - Develop data handling policies to ensure compliance with data privacy laws like GDPR and HIPAA.
- Ethical Review and Approval:
  - Set up a cross-functional ethics committee comprising data scientists, legal experts, and business leaders.
  - Define a process for ethical review and approval of new AI use cases, focusing on fairness, accountability, and transparency.

Industry Example

- **Healthcare**: A hospital exploring AI for medical image analysis establishes foundational governance by creating a data catalog using **Azure Purview** to document patient data and track its usage. An ethics committee is formed to review the project's ethical implications, ensuring patient privacy and compliance with HIPAA.

2.2. Stage 2: Intermediate Deployment – Implementing Formal Governance Structures

Overview

In the intermediate stage, Generative AI initiatives expand and begin to impact core business functions. With increased adoption, the risk of compliance issues and unintended consequences grows. Governance frameworks at this stage focus on formalizing policies, defining roles and responsibilities, and establishing structured compliance mechanisms. Organizations implement formal model governance frameworks, develop standardized documentation practices, and introduce explainability and transparency into their AI systems.

Governance Objectives at Stage 2

1. **Standardize Model Development and Validation Processes**: Create a standardized framework for model development, testing, and deployment to ensure consistent quality.
2. **Implement Robust Compliance Management**: Develop processes and tools to ensure adherence to industry regulations and internal policies.
3. **Enhance Model Explainability and Transparency**: Use explainability tools to provide visibility into AI decision-making and build trust with stakeholders.

Key Strategies

- **Model Governance Frameworks:**
  - Use tools like **MLflow**, **Azure Machine Learning Model Registry**, and **Google Vertex AI** to track model versions, monitor performance, and enforce governance policies.
  - Implement guidelines for model documentation, performance testing, and bias assessments.

- **Compliance and Risk Management:**
  - Use **AWS Config**, **Azure Policy**, or **Google Cloud Compliance Suite** to automate compliance checks and ensure models adhere to regulatory standards.
  - Establish a formal risk assessment process to evaluate new AI models before deployment.
- **Explainability and Transparency:**
  - Use **Google Explainable AI**, **Azure InterpretML**, or **AWS SageMaker Clarify** to provide insights into how models make decisions, making it easier to detect and mitigate biases.

Industry Example

- **Finance**: A bank scaling its AI-powered credit scoring system uses **AWS SageMaker Clarify** to detect potential biases in model predictions and **AWS Config** to ensure compliance with financial regulations. They standardize model documentation and validation processes, creating a comprehensive model registry for auditability and transparency.

2.3. Stage 3: Advanced Deployment – Establishing an AI Governance Center of Excellence

Overview

At this stage, AI becomes deeply embedded in mission-critical business processes, requiring a centralized governance structure. Organizations establish an AI Governance Center of Excellence (CoE) to coordinate governance activities across business units. The CoE defines enterprise-wide AI policies, manages risk, and promotes responsible AI practices. The emphasis is on creating a unified governance framework that supports scalability, consistency, and compliance across diverse AI initiatives.

Governance Objectives at Stage 3

1. **Centralize AI Governance and Oversight**: Establish a CoE to coordinate governance activities, define policies, and provide strategic direction.
2. **Implement Enterprise-Wide Governance Standards**: Develop a unified governance framework that ensures consistency across all AI initiatives.
3. **Automate Governance and Compliance Activities**: Use AI and automation to streamline governance processes, reducing manual effort and improving scalability.

Key Strategies

- **AI Governance CoE:**
  - Create a centralized AI Governance CoE that serves as a single point of accountability for all AI governance activities.
  - Define roles and responsibilities for data stewards, model validators, and compliance officers.
- **Unified Governance Framework:**
  - Develop standardized policies for data handling, model development, and deployment.

    - Implement a governance playbook that outlines best practices for AI development and deployment.
- **Automation of Governance Activities:**
    - Use tools like **AWS Control Tower**, **Azure Policy**, and **Google Cloud Org Policy** to automate policy enforcement, access management, and compliance checks.

Industry Example

- **Retail**: A global retailer establishes an AI Governance CoE to oversee its AI initiatives across different regions. The CoE develops a governance framework that includes standardized policies, automated compliance monitoring, and regular audits to ensure ethical and transparent use of AI technologies.

2.4. Stage 4: Mature AI Adoption – Embedding Governance into AI Operations

Overview

In the mature stage, governance is deeply integrated into the operational fabric of the organization. Governance processes are automated and continuously monitored, allowing the organization to adapt to changing regulations, business conditions, and technological advances. The focus is on creating a dynamic and adaptable governance framework that supports continuous improvement and scalability.

Governance Objectives at Stage 4

1. **Operationalize Governance Across the AI Lifecycle**: Embed governance into every aspect of AI operations, from data acquisition to model decommissioning.
2. **Implement Continuous Risk and Compliance Monitoring**: Develop real-time monitoring and risk management mechanisms to detect and respond to governance issues.
3. **Foster a Culture of Responsible AI**: Promote responsible AI practices across the organization, making them an integral part of the AI development process.

Key Strategies

- End-to-End AI Lifecycle Governance:
    - Use platforms like **Azure Machine Learning MLOps**, **AWS SageMaker MLOps**, and **Google Vertex AI Pipelines** to manage governance throughout the AI lifecycle.
    - Implement continuous integration and continuous deployment (CI/CD) pipelines with built-in governance controls.
- Real-Time Monitoring and Adaptability:
    - Use **AWS CloudWatch**, **Azure Monitor**, and **Google Cloud Monitoring** to track model performance and compliance in real time.
    - Implement adaptive governance mechanisms that adjust policies based on changing regulations or business needs.
- Creating a Responsible AI Culture:
    - Provide training and awareness programs on responsible AI practices.
    - Establish a feedback loop that encourages employees to report ethical concerns or governance issues.

Industry Example

- **Manufacturing**: A manufacturing firm uses **Google Vertex AI** to govern its AI-driven production optimization systems. They implement real-time compliance checks and adaptive governance mechanisms to ensure that the system remains aligned with safety standards and production goals, even as production requirements change dynamically.

3. Key Elements of an Effective AI Governance Framework

3.1. Data Governance

Effective data governance ensures that data used in AI models is reliable, secure, and ethically sourced. It includes policies for data quality, access control, lineage tracking, and privacy protection.

3.2. Model Governance

Model governance focuses on the lifecycle management of AI models, including documentation, validation, monitoring, and decommissioning. It ensures that models are transparent, explainable, and compliant.

3.3. Compliance and Risk Management

Compliance management involves adhering to legal and regulatory standards, while risk management focuses on identifying and mitigating potential risks associated with AI deployment.

3.4. Ethical Governance

Ethical governance addresses fairness, accountability, and transparency in AI systems. It includes developing ethical guidelines, conducting bias assessments, and ensuring that AI decisions align with societal values.

3.5. Accountability and Transparency

Clear accountability structures and transparent reporting mechanisms are essential for building trust in AI systems. Organizations must define roles and responsibilities and implement mechanisms for monitoring and auditing AI systems.

4. Conclusion

Effective governance is essential for advancing Generative AI maturity within the enterprise. As organizations progress through the stages of AI maturity, governance frameworks must evolve to address increasingly complex challenges. By establishing strong data governance, model governance, compliance management, and ethical oversight, organizations can ensure that their Generative AI systems are not only innovative but also responsible, compliant, and aligned with their business goals.

Chapter 13

Challenges with Generative AI Models and Strategies for Mitigation

Generative AI models, such as large-scale language models and architectures like Retrieval-Augmented Generation (RAG), have demonstrated groundbreaking capabilities in producing human-like text, creative content, and contextually rich responses. However, these models are not without limitations and inherent challenges that must be addressed before they can be reliably deployed in real-world scenarios. The complexities involved in building, fine-tuning, and deploying these models often result in issues such as hallucinations, biases, inability to handle long contexts, computational costs, and lack of interpretability.

This chapter provides a detailed overview of the challenges associated with Generative AI, practical solutions to address each challenge, and best practices for mitigating risks. It also highlights the tools and services provided by AWS, Azure, and Google Cloud to overcome these limitations, ensuring robust and trustworthy AI deployments.

1. Understanding Key Challenges with Generative AI Models

1.1. Hallucinations in Generative AI

Description: Hallucinations refer to instances where a generative model produces content that is factually incorrect, misleading, or fabricated. This happens when the model, despite being trained on large-scale datasets, generates outputs that are plausible-sounding but not grounded in any actual data.

- **Example**: A medical AI assistant might generate a diagnosis based on symptoms that sound realistic but have no grounding in medical science.
- **Impact**: Hallucinations can lead to misinformation, undermine user trust, and pose risks in sensitive applications like healthcare, legal advisory, and finance.

Causes:

- **Lack of Grounding**: Generative models like GPT-3 are typically trained using self-supervised learning without direct access to structured knowledge bases.
- **Model Overconfidence**: Generative models tend to produce confident-sounding answers even when they are unsure, leading to incorrect outputs.

Solution Strategies:

1. **Incorporate Retrieval-Augmented Generation (RAG):**
   - Implement RAG to provide the model with access to a structured knowledge base or external documents, grounding its responses in factual data.
   - Use tools like **Dense Passage Retrieval (DPR)** and **BM25** to find relevant documents and pass them to the generator as context.
2. **Fact-Checking Mechanisms:**
   - Use fact-checking tools and algorithms that automatically validate the generated content against trusted databases or sources.
   - Implement models like **TruthChecker** to cross-verify outputs for specific domains.
3. **Prompt Design and Controlled Generation:**
   - Use controlled generation techniques, such as reinforcement learning from human feedback (RLHF), to train models to generate outputs that are more aligned with factual data.
   - Implement context-specific prompts that guide the model to provide verifiable information.

Best Practices:

- Always combine generative models with a retrieval component for sensitive use cases.
- Regularly update the retrieval database to ensure the model has access to the latest information.

1.2. Data Biases and Ethical Concerns

Description: Generative AI models are trained on vast and diverse datasets, which often contain inherent biases, stereotypes, and inappropriate content. As a result, these biases are reflected in the model's outputs, leading to discriminatory, unethical, or biased content generation.

- **Example**: An AI model may generate biased job recommendations, favoring certain genders for specific roles.
- **Impact**: Biased outputs can lead to negative societal implications, discrimination, and compliance issues, especially in regulated industries like finance, HR, and healthcare.

Causes:

- **Training Data Bias**: Datasets used for training may include skewed or biased representations of gender, race, or culture.
- **Lack of Fairness Metrics**: Traditional training processes do not always include fairness metrics or optimization strategies.

Solution Strategies:

1. **Bias Mitigation During Pre-Processing:**
    - Use data preprocessing techniques like **re-sampling**, **re-weighting**, and **counterfactual data augmentation** to create balanced datasets.
    - Implement **domain-specific data cleaning** to remove or minimize biased content.

2. **Post-Training Bias Detection:**
    - Use tools like **AWS SageMaker Clarify**, **Google Cloud AI Explanations**, or **Azure Fairness SDK** to detect and quantify biases in the model's predictions.
    - Implement evaluation metrics like **equal opportunity difference**, **demographic parity**, and **disparate impact ratio**.
3. **Fairness-Constrained Training:**
    - Incorporate fairness constraints directly into the loss function to minimize biased outputs.
    - Use adversarial training to penalize biased predictions and optimize for fairness.

Best Practices:

- Perform a thorough **fairness assessment** before deploying models.
- Regularly audit models for bias using tools like **AI Fairness 360** or **Fairlearn**.

1.3. Handling Long Contexts and Multi-Turn Conversations

Description: Most generative models have a fixed context window, which limits their ability to handle long documents, multi-turn conversations, or complex tasks requiring memory of previous interactions.

- **Example**: A chatbot may lose track of context in a long customer support interaction, resulting in repeated or irrelevant responses.
- **Impact**: Inability to handle long contexts reduces the effectiveness of the model in scenarios like document summarization, long-form dialogue systems, and contract analysis.

Causes:

- **Fixed Context Length**: Models like GPT-3 are limited by a fixed number of tokens, making them unsuitable for long inputs.
- **Lack of Memory Mechanisms**: Most transformers lack the capability to retain information beyond the current input.

Solution Strategies:

1. **Use Extended Context Models:**
    - Implement models like **Longformer**, **Reformer**, or **Performer**, which use sparse attention mechanisms to extend the context window.

- Use memory-augmented architectures like **Compressive Transformers** to store and access long-term context.

2. **Context Chaining and Chunking:**
   - Break long inputs into smaller chunks and process them sequentially.
   - Implement memory mechanisms that allow the model to refer back to previous chunks.
3. **Hierarchical Attention Mechanisms:**
   - Use hierarchical transformers to process long documents in a multi-scale fashion, capturing both local and global context.

Best Practices:

- Choose models based on the length of the context required for your specific application.
- Experiment with hierarchical and memory-augmented models for improved long-context understanding.

1.4. High Computational Costs

Description: Generative models are resource-intensive, requiring large amounts of compute power, memory, and storage for both training and inference. This becomes a major bottleneck when scaling models for production use or fine-tuning them on custom data.

- **Example**: Training GPT-3 requires thousands of GPU hours, making it infeasible for most organizations.
- **Impact**: High costs can limit accessibility and scalability, making it challenging to deploy these models in cost-sensitive environments.

Causes:

- **Large Parameter Count**: Modern models often have billions of parameters, increasing computational requirements.
- **Complex Training Workflows**: Distributed training, gradient accumulation, and checkpointing add to the overhead.

Solution Strategies:

1. **Model Optimization Techniques:**
   - Use **model pruning**, **quantization**, and **distillation** to reduce model size and computation.
   - Implement **layer-wise adaptive learning rates** to accelerate convergence.
2. **Use Pre-trained and Distilled Models:**
   - Start with pre-trained models from **Hugging Face** or **Azure Model Zoo** and fine-tune them instead of training from scratch.
   - Use **distilled versions** (e.g., DistilBERT) for lightweight deployments.
3. **Optimize Training with Distributed Frameworks:**
   - Use **DeepSpeed** (Azure), **Horovod** (AWS), or **TPU Pods** (Google Cloud) to optimize training across multiple GPUs or TPUs.

Best Practices:

- Implement mixed-precision training using **FP16** for faster computation.
- Use **gradient checkpointing** to reduce memory usage during training.

2. Tools and Services for Mitigating Generative AI Challenges

Cloud platforms like AWS, Azure, and Google Cloud offer a range of tools to mitigate challenges associated with Generative AI models.

AWS Tools

1. **SageMaker Clarify**: Detects bias and provides interpretability metrics.
2. **SageMaker Debugger**: Monitors training in real time, identifying inefficiencies.
3. **AWS Comprehend**: Provides NLP services with built-in content moderation and entity recognition.

Azure Tools

1. **Azure Machine Learning Fairness SDK**: Offers fairness and bias detection metrics.
2. **Azure Cognitive Search**: Integrates with RAG systems to provide semantic search capabilities.
3. **Azure Synapse Analytics**: Facilitates large-scale data processing for complex AI models.

Google Cloud Tools

1. **Explainable AI**: Provides interpretability through feature attributions.
2. **Vertex AI Model Monitoring**: Detects drift and monitors model health.
3. **Google Cloud Natural Language API**: Pre-built NLP models for entity recognition and sentiment analysis.

3. Conclusion

Successfully deploying Generative AI models requires a deep understanding of their limitations and implementing robust strategies to address common challenges. By leveraging advanced techniques and cloud-based tools, organizations can build AI solutions that are reliable, scalable, and aligned with ethical standards, paving the way for broader adoption of Generative AI in diverse industries.

Chapter 14

Responsible AI — Building Trust and Accountability

With the proliferation of AI technologies across industries, ensuring that AI is developed and deployed in an ethical, fair, and responsible manner is now more critical than ever. As AI systems influence decision-making processes in sectors like finance, healthcare, law enforcement, and beyond, they hold the power to shape individual lives and society at large. Responsible AI is not only a moral obligation but also a business necessity to avoid unintended consequences, ensure compliance with regulations, and maintain trust among customers and stakeholders.

In this chapter, we will delve into the relevance and necessity of Responsible AI, discuss its implications for society and businesses, and outline a comprehensive governance framework to guide organizations. We will also explore the specific tools and services offered by major cloud providers—AWS, Google Cloud, and Microsoft Azure—to support the adoption of Responsible AI.

1. Why Responsible AI is Relevant

AI technologies have become ubiquitous, powering everything from recommendation systems to autonomous vehicles and predictive analytics. As AI systems become more integrated into daily life, the stakes for ensuring their ethical and responsible use continue to rise. Here are key reasons why Responsible AI is relevant and critical:

1.1. Ethical Considerations in AI Development

AI systems can reinforce biases present in historical data or the assumptions of their developers. Without intentional safeguards, AI models can inadvertently perpetuate stereotypes or make unfair decisions. For example:

- **Hiring Systems**: AI models used in hiring can favor certain demographics if trained on biased data.
- **Law Enforcement**: Predictive policing models can lead to over-policing of minority communities based on historical crime data.

Implementing Responsible AI practices ensures that ethical considerations are embedded throughout the AI lifecycle, from data collection to model deployment.

1.2. Regulatory and Legal Compliance

As AI adoption grows, so does regulatory scrutiny. Governments and international bodies are introducing regulations to govern AI use, making compliance a business imperative.

- **European Union's AI Act**: Proposes a risk-based approach to regulating AI, requiring organizations to conduct impact assessments and implement risk management frameworks.
- **General Data Protection Regulation (GDPR)**: Requires transparency, fairness, and accountability when using AI to process personal data. Non-compliance can result in fines of up to 4% of global annual revenue.

Organizations need to ensure that their AI systems meet regulatory requirements to avoid legal liabilities and operational disruptions.

1.3. Business and Reputational Risks

Deploying AI without adequate governance can lead to decisions that damage an organization's reputation. Negative public perception of AI technologies due to ethical lapses can result in loss of customer trust, reduced market share, and even boycotts.

- **Case Study**: A large technology company faced backlash when its facial recognition system was found to perform poorly on women and individuals with darker skin tones. The controversy led to calls for stricter regulations and impacted the company's public image.

1.4. Societal Impact

AI has the potential to bring about positive societal change, such as improving healthcare diagnostics or optimizing supply chains to reduce waste. However, if not managed responsibly, AI can exacerbate existing inequalities or create new forms of harm.

- **Example**: AI systems that allocate resources or recommend treatments must be designed to avoid discriminating against marginalized groups.

Responsible AI ensures that the benefits of AI are shared equitably and that potential harms are minimized.

2. Necessity of Responsible AI: Key Considerations

Responsible AI is necessary to build AI systems that are ethical, fair, and aligned with societal values. The following considerations highlight why it is crucial for enterprises to prioritize Responsible AI:

2.1. Mitigating Bias and Ensuring Fairness

AI systems trained on historical data may inadvertently perpetuate existing biases, leading to unfair outcomes in critical areas like lending, hiring, and law enforcement.

- **Fairness**: AI models should be designed to treat all individuals fairly, without discrimination based on race, gender, age, or other protected attributes.
- **Bias Mitigation Techniques**: Use techniques like re-weighting, adversarial debiasing, and fairness constraints during training to reduce bias.

2.2. Building Trust and Transparency

Trust is a cornerstone of AI adoption. Users are more likely to accept AI systems if they understand how decisions are made and can challenge those decisions if necessary.

- **Transparency**: Implement explainable AI techniques to make model decisions more interpretable.
- **Explainability**: Use frameworks like LIME (Local Interpretable Model-Agnostic Explanations) or SHAP (SHapley Additive exPlanations) to provide insights into model predictions.

2.3. Ensuring Accountability and Traceability

AI systems should have clear lines of accountability. If an AI system causes harm, it should be possible to trace the root cause and hold the appropriate stakeholders accountable.

- **Documentation**: Maintain thorough documentation of model development, including datasets, design decisions, and testing results.
- **Audit Trails**: Implement audit trails to track model outputs and decision-making processes over time.

2.4. Protecting Privacy and Data Rights

AI systems often require large amounts of personal data for training and operation. Protecting user privacy is critical to maintaining trust and adhering to data protection regulations.

- **Techniques**: Use differential privacy, federated learning, and data anonymization to protect sensitive information.
- **Compliance**: Ensure that data collection and usage comply with regulations like GDPR, CCPA, and HIPAA.

3. Governance Framework for Responsible AI

Implementing a governance framework is essential to ensuring that AI systems are developed, deployed, and monitored in a responsible manner. Below are key components of a Responsible AI governance framework:

3.1. AI Governance Committees

Establish cross-functional committees that include representatives from technical, legal, and business teams. These committees should oversee AI projects, review model design, and ensure alignment with ethical guidelines.

- **Role**: Set policies, review high-risk AI models, and provide guidance on ethical considerations.

- **Structure**: Include roles such as Chief AI Officer, AI Ethics Officers, and domain experts.

3.2. Ethical Review Boards

Create ethical review boards to evaluate AI systems for potential harm, fairness, and societal impact. These boards should have the authority to halt projects if ethical concerns are identified.

- **Role**: Conduct impact assessments, provide recommendations for mitigating risks, and ensure compliance with ethical standards.

3.3. AI Model Risk Management Framework

Implement a model risk management framework that includes guidelines for model validation, monitoring, and auditing. The framework should address:

- **Model Validation**: Test models for performance, bias, and robustness.
- **Model Monitoring**: Continuously monitor models for drift, bias, and unintended behaviors.
- **Model Auditing**: Conduct regular audits to ensure compliance with Responsible AI principles.

3.4. Accountability Mechanisms

Assign clear accountability for AI projects. Create roles such as AI Stewards who are responsible for overseeing ethical considerations at each stage of the AI lifecycle.

3.5. Employee Training Programs

Provide ongoing training for employees involved in AI development. Training should cover topics like fairness, bias mitigation, transparency, and privacy.

- **Content**: Include practical workshops, case studies, and guidelines on implementing Responsible AI practices.

4. Tools and Services for Implementing Responsible AI

AWS, Google Cloud, and Microsoft Azure provide various tools and services to support Responsible AI. Below is an overview of what each cloud provider offers:

4.1. Amazon Web Services (AWS)

AWS provides several tools and services to support Responsible AI practices:

- **Amazon SageMaker Clarify**: Detects bias in datasets and models, providing explainability reports to help interpret model predictions.
- **Amazon SageMaker Model Monitor**: Continuously monitors models in production for bias and performance issues.
- **Amazon Macie**: Uses machine learning to detect sensitive data and ensure compliance with data privacy regulations.

- **AWS Identity and Access Management (IAM)**: Helps implement role-based access control to protect sensitive AI assets.

4.2. Google Cloud

Google Cloud offers a suite of tools designed to promote fairness, explainability, and privacy:

- **What-If Tool**: A visual tool for analyzing machine learning models, detecting bias, and testing model performance on different subpopulations.
- **AI Explanations**: Provides insights into model predictions, helping to understand how input features influence model outputs.
- **Data Privacy Tools**: Google's Differential Privacy libraries enable the development of privacy-preserving AI systems.
- **Vertex AI Model Monitoring**: Continuously monitors model performance and fairness.

4.3. Microsoft Azure

Microsoft Azure emphasizes transparency and accountability through its Responsible AI services:

- **Azure Machine Learning Fairness Dashboard**: Identifies and mitigates bias in machine learning models.
- **Azure AI Explainability Toolkit**: Provides tools like SHAP and interpretML to make models more transparent.
- **Azure Security Center**: Protects AI workloads from threats, ensuring that AI systems are secure and compliant.
- **Responsible AI Dashboard**: Provides an integrated view of model fairness, explainability, error analysis, and data exploration.

5. Global Developments in Responsible AI Policy

Global developments in responsible artificial intelligence (AI) policy have accelerated, with nations and international organizations implementing frameworks to ensure ethical AI development and deployment.

United States: In October 2023, President Biden issued an executive order establishing guidelines for AI use within national security, emphasizing adherence to democratic values and human rights. This directive aims to prevent misuse, such as infringing on free speech or bypassing nuclear controls, and underscores the necessity of human oversight and accountability in AI applications.

European Union: The EU is set to implement the AI Act, the first comprehensive AI regulation, in December 2024. This legislation emphasizes trust, transparency, and accountability, imposing strict transparency requirements on high-risk AI systems and restricting government use of real-time biometric surveillance to specific serious crime scenarios. The AI Act is expected to influence global standards, with full application beginning in 2026.

United Nations: The UN's AI advisory body released a report in September 2024 with seven recommendations for governing AI, addressing risks and gaps in international governance. Key proposals include establishing a panel to provide impartial AI knowledge, launching a policy dialogue on AI governance, and creating an AI standards exchange and a global capacity development network. The advisory body also suggests forming a global AI fund to address capacity gaps and a global data framework to ensure transparency. A small AI office is proposed to coordinate these initiatives.

Australia: Voluntary AI Safety Standards and Mandatory Safeguards

The Australian Government has introduced voluntary AI safety standards to guide organizations in the responsible development and deployment of AI technologies. These standards focus on risk management, human oversight, and transparency. Additionally, the government is considering implementing mandatory safeguards for high-risk AI applications to ensure public safety and trust.

On 1 September 2024, the Australian Government implemented the "Policy for the Responsible Use of AI in Government." This policy provides a unified approach for government agencies to engage with AI confidently, safely, and responsibly. It emphasizes transparency, governance, and risk assurance to strengthen public trust. The policy is designed to evolve over time, adapting to technological advancements and changing community expectations

Global Partnerships: The Global Partnership on AI (GPAI), an international initiative launched in June 2020, promotes responsible AI use that respects human rights and democratic values. With 25 member countries, including Australia, Canada, France, Germany, India, Japan, and the United States, GPAI brings together experts from industry, government, civil society, and academia to advance research and pilot projects on AI priorities.

Industry Initiatives: The World Economic Forum's AI Governance Alliance, established in June 2023, focuses on responsible generative AI development. The alliance brings together governments, businesses, and experts to shape responsible AI development applications and governance, ensuring equitable distribution and enhanced access to AI technologies worldwide.

These initiatives reflect a global consensus on the need for responsible AI governance, balancing innovation with ethical considerations to mitigate risks and ensure AI technologies benefit society.

6. Conclusion

Responsible AI is essential for building trust, ensuring compliance, and minimizing harm as AI systems become integral to business operations and society. By implementing a robust governance framework and leveraging cloud tools and services, organizations can build AI systems that are not only powerful but also ethical and aligned with societal values. As the AI landscape evolves, enterprises must remain vigilant and proactive in addressing the ethical challenges posed by AI technologies to ensure that AI serves humanity in a positive and beneficial manner.

Governments worldwide, are increasingly recognizing the importance of responsible AI to safeguard public interests while promoting innovation. Initiatives such as the Australian Government's "Policy for the Responsible Use of AI in Government" and the "National Framework for the Assurance of Artificial Intelligence" exemplify proactive measures to establish ethical standards and transparent governance structures. Globally, policies like the U.S. executive orders on AI use, the EU's forthcoming AI Act, and the United Nations' advisory recommendations further underscore the global commitment to ethical AI development. These initiatives reflect a collective approach to address AI's complex ethical, safety, and governance challenges, ensuring that AI technologies are aligned with societal values, support public trust, and are used responsibly across industries. As these frameworks continue to evolve, they will play a pivotal role in fostering AI that is safe, transparent, and beneficial, serving as a model for future advancements in technology policy.

Chapter 15

Misuse of Generative AI: Threats, Risks, and Enterprise Protection Strategies

Generative AI (GenAI) has unlocked transformative capabilities across industries, enabling organizations to automate content creation, enhance customer interactions, and gain new insights. However, the same powerful technologies that drive innovation can also be misused for malicious purposes. This chapter explores the dark side of Generative AI, highlighting the key misuse scenarios, potential risks, and how enterprises can safeguard themselves against these emerging threats.

The misuse of Generative AI spans a wide spectrum, from creating deepfake videos and spreading disinformation to intellectual property (IP) theft, social engineering, and data manipulation. These malicious applications pose significant risks, including reputational damage, regulatory penalties, financial loss, and ethical dilemmas. To address these challenges, organizations must implement comprehensive governance frameworks, enforce strict security measures, and continuously monitor for anomalous behaviors.

The chapter breaks down the most common and dangerous misuse scenarios, such as misinformation campaigns, data poisoning, AI-generated phishing attacks, and the creation of AI-powered malware. It provides real-world examples to illustrate how these threats manifest in different industries, causing harm to both businesses and society. For each misuse scenario, the chapter outlines specific protection strategies, from implementing ethical AI guidelines to leveraging AI-powered detection tools.

By understanding the potential misuse of Generative AI and adopting robust defense strategies, enterprises can protect themselves and their stakeholders from the negative consequences of these technologies. Through a combination of proactive governance, technical controls, and ethical considerations, organizations can harness the power of Generative AI while mitigating its risks.

1. Misinformation and Disinformation

One of the most alarming misuses of Generative AI is its ability to produce realistic but entirely fabricated content, such as deepfake videos, fake news articles, or AI-generated social media posts. Misinformation and disinformation campaigns can be launched at scale, influencing public opinion, sowing discord, and causing financial and reputational damage to individuals, corporations, and governments.

Examples of Misuse:

1. **Deepfake Videos of Public Figures:**
   - Deepfakes are hyper-realistic AI-generated videos that manipulate the likeness and voice of individuals. This can be used to depict politicians, business leaders, or celebrities saying or doing things they never did, potentially altering public perception or stock prices.
   - **Use Case**: In 2020, a deepfake of a popular CEO announcing a fraudulent partnership caused a temporary spike in the company's stock price before the misinformation was debunked.
2. **Fabricated News Articles:**
   - Generative AI models like GPT-3 can create entire articles or social media threads that appear to be legitimate news, making it difficult to distinguish between truth and falsehood.
   - **Use Case**: In 2019, AI-generated articles were used in a coordinated disinformation campaign that spread false information about a pharmaceutical company, causing a sharp drop in its stock value.
3. **Social Media Manipulation:**
   - AI models can generate thousands of fake comments, tweets, or posts that promote false narratives or target specific groups, manipulating opinions and spreading harmful content.
   - **Use Case**: In a coordinated misinformation campaign, fake social media accounts generated by GenAI were used to post thousands of comments supporting a particular political viewpoint, influencing election results in a small country.

Impact on Enterprises:

- **Reputation Damage**: False narratives can harm the reputation of companies or executives.
- **Financial Loss**: Misinformation can impact stock prices or cause loss of consumer trust.
- **Operational Disruption**: Misinformation campaigns can disrupt operations by instigating public backlash or boycotts.

Protection Strategies:

1. **AI-Based Detection and Mitigation Tools:**
   - Use tools like Microsoft's **Video Authenticator** to detect deepfakes and Adobe's **Content Authenticity Initiative** to verify the authenticity of images and videos.
2. **Content Authentication and Provenance Technologies:**
   - Implement blockchain-based content authentication solutions to establish the authenticity and source of digital content.

3. **Collaboration with Fact-Checking Networks:**
   - Partner with external fact-checking organizations to verify content and establish a rapid-response protocol for misinformation incidents.

2. Intellectual Property (IP) Theft and Data Misappropriation

Overview

Generative AI models can generate content—whether text, images, music, or code—that closely resembles or even duplicates copyrighted material. This issue arises when models are trained on large datasets that may contain copyrighted content without proper licensing. IP theft and data misappropriation can expose organizations to legal liabilities and reputational damage.

Examples of Misuse:

1. **Unauthorized Generation of Copyrighted Content:**
   - Generative AI models trained on copyrighted books, music, or code can produce derivative works that infringe on IP rights.
   - **Use Case**: An AI-based art generator was accused of producing artwork that closely resembled the styles of living artists, leading to a lawsuit over copyright infringement.
2. **Reproducing Proprietary Source Code:**
   - AI models like Codex or GPT-3 can generate code snippets that may unintentionally include proprietary code, violating IP laws.
   - **Use Case**: A technology company found that an AI-based code generator had reproduced proprietary code from its internal repositories, potentially exposing critical IP to external users.
3. **Generating Derivative Content Without Attribution:**
   - AI models can create derivative text or designs without crediting the original creators, raising ethical and legal concerns.
   - **Use Case**: A startup used a text generation model to rewrite articles from a competitor's website, resulting in legal action for plagiarism and unfair use of IP.

Impact on Enterprises:

- **Legal Liabilities**: Organizations may face lawsuits for copyright infringement or IP theft.
- **Reputational Damage**: Misuse of copyrighted materials can damage a brand's reputation and ethical standing.
- **Financial Penalties**: Regulatory fines and compensation payments may be required for IP violations.

Protection Strategies:

1. **IP Compliance Checks:**
   - Implement tools to automatically check generated content for IP violations against publicly available databases of copyrighted material.

2. **Strict Data Governance Policies:**
    - Ensure that AI models are only trained on legally permissible and properly licensed datasets. Use contracts and agreements to govern data usage rights.
3. **Implement Usage Policies and AI Model Audits:**
    - Establish strict policies on how generated content can be used, and conduct regular audits of model outputs for compliance.

3. Social Engineering and Phishing Attacks

Overview

Generative AI models can automate and personalize phishing emails or social engineering attacks at scale. By mimicking the communication style of known individuals, GenAI can deceive victims into revealing sensitive information or performing unauthorized actions.

Examples of Misuse:

1. **Business Email Compromise (BEC) Scams:**
    - AI models can generate highly personalized emails that appear to come from a company's CEO or CFO, instructing employees to initiate fraudulent wire transfers.
    - **Use Case**: A German energy company lost €220,000 after an AI-generated voice mimicking the CEO instructed a finance executive to make a wire transfer to a fraudulent account.
2. **Spear Phishing Emails:**
    - GenAI can create spear phishing emails that incorporate detailed information about the target, making the emails appear highly credible.
    - **Use Case**: A cybersecurity firm identified a campaign where AI-generated emails were used to target financial institutions, leading to compromised employee credentials.
3. **Automated Social Engineering Attacks:**
    - AI models can generate fake LinkedIn messages, direct messages on social media, or other communications that exploit trust and familiarity.
    - **Use Case**: An attacker used a GenAI model to impersonate a prominent cybersecurity expert, tricking several individuals into downloading malware under the guise of security updates.

Impact on Enterprises:

- **Data Breaches**: Employees tricked by AI-generated phishing attempts may unwittingly expose sensitive data.
- **Financial Losses**: Organizations may lose funds through fraudulent transactions.
- **Operational Disruption**: Successful attacks can lead to compromised systems and costly recovery efforts.

Protection Strategies:

1. **AI-Based Email Security Solutions:**

- Deploy advanced email filtering solutions that use machine learning to identify and flag suspicious patterns in emails.

2. **Implement Strong Authentication Mechanisms:**
   - Use multi-factor authentication (MFA) for all sensitive transactions and communications to mitigate the impact of successful phishing attacks.
3. **Regular Security Awareness Training:**
   - Educate employees about AI-generated phishing threats and train them to identify and report suspicious communications.

4. Data Poisoning and Model Manipulation

Overview

Data poisoning involves injecting malicious or biased data into training datasets to corrupt the AI model's outputs. Model manipulation can result in models making incorrect decisions, generating harmful content, or amplifying biases. These attacks can be difficult to detect and have long-lasting effects on model integrity.

Examples of Misuse:

1. **Corrupting Financial Models:**
   - Attackers may introduce biased data into financial models, causing them to recommend poor investment decisions or favor specific outcomes.
   - **Use Case**: A hedge fund experienced significant losses after an attacker successfully poisoned its trading model's training data, leading to incorrect stock predictions.
2. **Altering Healthcare AI Models:**
   - Malicious actors can manipulate training data to cause healthcare models to misdiagnose conditions or recommend harmful treatments.
   - **Use Case**: A hospital's diagnostic model was found to misclassify certain rare diseases after its training data was manipulated by an insider.
3. **Embedding Backdoors in AI Models:**
   - Attackers can introduce hidden triggers during training that cause the model to behave unexpectedly when specific inputs are provided.
   - **Use Case**: A cybersecurity firm identified a language model with a backdoor that generated specific outputs when given seemingly benign keywords, leading to sensitive information leakage.

Impact on Enterprises:

- **Model Degradation**: Poisoned models produce unreliable outputs, leading to poor business decisions.
- **Compliance Violations**: Manipulated models may inadvertently produce biased or discriminatory outcomes, leading to legal consequences.
- **Operational Risks**: Models that behave unpredictably can disrupt business operations.

Protection Strategies:

1. **Robust Data Validation Processes:**
   - Use automated tools to validate and sanitize training data before it is used in model development.
2. **Adversarial Testing and Robustness Assessments:**
   - Conduct regular adversarial testing to identify vulnerabilities and ensure model robustness against data manipulation.
3. **Real-Time Model Monitoring and Retraining:**
   - Continuously monitor models for anomalies and retrain them if unusual patterns are detected.

How Enterprises Can Protect Against Misuse of Generative AI

Overview

As Generative AI continues to evolve, its misuse poses serious risks for enterprises, including data security breaches, reputational damage, financial losses, and regulatory non-compliance. This chapter outlines comprehensive strategies and best practices for safeguarding enterprises against these threats. It covers the key components of AI governance, ethical AI guidelines, security frameworks, and risk management measures that organizations must implement to protect themselves from the potential misuse of Generative AI.

This chapter also provides detailed use cases from various industries to illustrate how these strategies can be effectively applied in real-world scenarios. By following the outlined recommendations, organizations can establish robust defenses to ensure that Generative AI technologies are used responsibly and securely.

1. Establishing a Comprehensive AI Governance Framework

Overview

A robust AI governance framework provides the foundation for responsible and secure AI deployment. It defines the policies, standards, and controls that govern the use of Generative AI across the enterprise, ensuring alignment with organizational goals and compliance with regulatory requirements. A governance framework should cover all aspects of AI operations, including data management, model development, ethical considerations, and risk management.

Key Strategies for Implementing AI Governance

1. **Create a Cross-Functional AI Governance Board:**
   - Establish an AI Governance Board that includes representatives from data science, IT, legal, compliance, and business units. This board is responsible for setting governance policies, reviewing AI use cases, and ensuring compliance.
   - **Example**: A large financial institution formed a cross-functional AI governance board to oversee its AI initiatives, ensuring that models used for customer credit scoring were compliant with internal and external regulations.

2. **Develop AI Usage Policies and Guidelines:**
   - Define clear policies for acceptable and unacceptable uses of Generative AI. These policies should specify the types of data that can be used, ethical considerations, and usage restrictions for sensitive applications like surveillance or social profiling.
   - **Example**: A healthcare company developed strict guidelines for using AI in patient diagnosis, prohibiting the use of Generative AI models for making treatment recommendations without human oversight.
3. **Implement Model and Data Governance Processes:**
   - Establish formal processes for data quality assurance, model documentation, version control, and performance monitoring. Use tools like **MLflow**, **Azure Machine Learning MLOps**, or **Google Vertex AI** to manage the model lifecycle and enforce governance standards.
   - **Example**: A global retailer uses **MLflow** to track model versions and ensure that every AI model deployed for sales forecasting meets governance standards for accuracy and bias.
4. **Create a Governance Playbook for AI Projects:**
   - Develop a playbook that outlines best practices for AI project management, including data handling procedures, model evaluation criteria, and ethical considerations. This playbook should serve as a reference for all AI stakeholders.
   - **Example**: A multinational company created an AI governance playbook that provides step-by-step guidance on how to assess AI models for bias, fairness, and compliance before deployment.

Impact of AI Governance

An effective AI governance framework minimizes the risk of misuse, ensures regulatory compliance, and fosters a culture of responsible AI across the organization.

2. Implementing Ethical AI Guidelines

Overview

Ethical AI guidelines help organizations align AI development and usage with societal values and organizational principles. These guidelines address issues such as fairness, accountability, transparency, and respect for privacy. Implementing ethical AI guidelines involves establishing a framework for ethical decision-making and providing guidance for evaluating AI projects from an ethical standpoint.

Key Strategies for Implementing Ethical AI Guidelines

1. **Define a Code of Ethics for AI:**
   - Establish a formal Code of Ethics for AI that outlines the organization's commitment to ethical AI practices. This code should address issues like fairness, bias mitigation, and respect for human rights.
   - **Example**: A major technology firm introduced a Code of Ethics for AI that prohibits the use of its AI models for mass surveillance or applications that infringe on civil liberties.

2. **Create an AI Ethics Committee:**
   - Set up an ethics committee to review AI use cases, assess potential ethical risks, and provide recommendations for ethically sensitive projects. This committee should include representatives from diverse backgrounds to ensure a holistic perspective on ethical issues.
   - **Example**: An insurance company established an AI ethics committee to review models used for policy underwriting, ensuring that they do not discriminate based on race, gender, or socioeconomic status.
3. **Develop an Ethical Impact Assessment Process:**
   - Implement an ethical impact assessment (EIA) process for all AI projects. The EIA should evaluate the potential ethical implications of the AI system, including issues like fairness, bias, and transparency.
   - **Example**: A healthcare startup developed an EIA process to assess the ethical impact of its AI-powered diagnostic tools, ensuring that they do not amplify biases in healthcare access or treatment.
4. **Provide Training on Ethical AI Practices:**
   - Offer training programs for data scientists, engineers, and business leaders on the principles of ethical AI. Training should cover topics such as bias detection, data privacy, and responsible AI design.
   - **Example**: A financial services firm created a mandatory training program for its AI development teams, focusing on bias detection and mitigation techniques to ensure fair lending practices.

Impact of Ethical AI Guidelines

Implementing ethical AI guidelines helps prevent unintended consequences, builds public trust, and ensures that AI initiatives align with the organization's values and societal expectations.

3. Strengthening AI Security Measures

Overview

AI security measures are critical for protecting AI models, data, and systems from cyber threats such as data poisoning, model manipulation, and adversarial attacks. Enterprises must implement a multi-layered security approach that includes data validation, model hardening, and continuous monitoring to safeguard against both internal and external threats.

Key Strategies for Strengthening AI Security

1. **Implement Data Security and Access Control Mechanisms:**
   - Use encryption, access controls, and secure data storage to protect training data and sensitive information. Implement tools like **Azure Key Vault**, **AWS KMS**, or **Google Cloud KMS** for key management and data encryption.
   - **Example**: A pharmaceutical company uses **Azure Key Vault** to secure patient data used in AI models, ensuring that only authorized personnel have access to sensitive information.

2. **Conduct Adversarial Testing and Robustness Assessments:**
    - Perform adversarial testing to identify vulnerabilities in AI models and assess their robustness against attacks. Use tools like the **Adversarial Robustness Toolbox** to simulate attack scenarios and evaluate model resilience.
    - **Example**: A bank uses the **Adversarial Robustness Toolbox** to test its fraud detection models against adversarial attacks, ensuring that the models are robust and resistant to manipulation.
3. **Implement Real-Time Model Monitoring:**
    - Set up real-time monitoring to detect anomalies in model behavior that could indicate malicious activity or data drift. Use tools like **AWS CloudWatch**, **Azure Monitor**, or **Google Cloud Monitoring** to monitor model performance and security.
    - **Example**: A manufacturing company implemented **AWS CloudWatch** to monitor its predictive maintenance models in real-time, enabling it to quickly identify and respond to potential data manipulation attempts.
4. **Develop Response Plans for AI Security Incidents:**
    - Create response plans for AI security incidents, outlining the steps to take in the event of a data breach, model manipulation, or adversarial attack. Ensure that the response plan includes communication protocols and recovery procedures.
    - **Example**: A telecom company developed an AI security incident response plan to address potential data poisoning attacks on its customer service chatbots, ensuring quick containment and recovery.

Impact of AI Security Measures

Strengthening AI security helps protect AI systems from malicious attacks, ensures data integrity, and maintains the trustworthiness of AI outputs.

4. Implementing Continuous Monitoring and Auditing

Overview

Continuous monitoring and auditing are essential for maintaining the security, compliance, and integrity of AI models over time. Regular audits help identify potential vulnerabilities, ensure adherence to policies, and provide transparency into AI operations. Monitoring tools can detect anomalies in real-time, enabling rapid response to emerging risks.

Key Strategies for Continuous Monitoring and Auditing

1. **Set Up Automated Monitoring for AI Models:**
    - Implement automated monitoring solutions to track model performance, data inputs, and outputs. Use tools like **AWS SageMaker Model Monitor**, **Azure Machine Learning Monitor**, or **Google Vertex AI Model Monitoring**.
    - **Example**: A logistics company uses **Azure Machine Learning Monitor** to track the performance of its demand forecasting models, enabling it to detect and address model drift in real-time.

2. **Conduct Regular AI Model Audits:**
    - Perform regular audits of AI models to assess compliance with governance policies, ethical guidelines, and security standards. Document the findings and implement corrective actions as needed.
    - **Example**: A global bank conducts quarterly audits of its AI models used for loan approvals to ensure they remain compliant with fairness and non-discrimination policies.
3. **Implement Real-Time Anomaly Detection:**
    - Use anomaly detection tools to identify unusual patterns in model behavior or data inputs that could indicate potential misuse or security breaches.
    - **Example**: An e-commerce company uses Google Cloud's anomaly detection tools to monitor its recommendation systems for unusual activity, ensuring that models are not manipulated to favor specific products.

Impact of Continuous Monitoring and Auditing

Continuous monitoring and auditing ensure that AI systems remain secure, compliant, and aligned with organizational goals, even as conditions change over time.

5. Establishing Clear Accountability and Governance Structures

Overview

Establishing clear accountability and governance structures is essential for ensuring that Generative AI systems are managed responsibly and securely. Accountability defines who is responsible for each aspect of AI governance, from data handling to model deployment and ethical oversight. This structure ensures that there is clarity around roles and responsibilities, enabling effective management and rapid response to potential risks or misuse scenarios.

Key Strategies for Establishing Accountability and Governance Structures

1. **Define Roles and Responsibilities for AI Governance**
    - Clearly define the roles and responsibilities for all stakeholders involved in AI development and governance, including data scientists, IT staff, legal teams, compliance officers, and business leaders. Create a responsibility matrix that outlines who is accountable, responsible, consulted, and informed (RACI) for each AI-related activity.
    - **Example**: A global bank defined a clear governance structure with specific roles for data stewards, model owners, compliance officers, and business leaders, ensuring that every step of the AI lifecycle—from data acquisition to model decommissioning—had an assigned owner.
2. **Establish an AI Risk Management Function**
    - Create a dedicated AI risk management team responsible for assessing, managing, and mitigating risks associated with AI deployments. This team should work closely with the AI Governance Board to identify potential threats and develop response strategies.

- **Example**: A multinational insurance company established an AI risk management team to evaluate the potential risks of using AI models for policy underwriting, including fairness, discrimination, and compliance risks.

3. **Set Up a Centralized AI Governance Hub**
   - Establish a centralized governance hub to coordinate AI governance activities across departments. This hub should maintain a repository of AI policies, track compliance, and serve as a single point of contact for AI governance inquiries and escalations.
   - **Example**: A technology company set up a centralized AI governance hub that oversees AI projects across its business units, ensuring that all models adhere to corporate governance standards and regulatory requirements.
4. **Create a Clear Escalation Path for AI Incidents**
   - Define a formal escalation path for AI incidents, outlining the steps to take in the event of a security breach, ethical concern, or compliance issue. Ensure that there are clear protocols for reporting and addressing incidents promptly.
   - **Example**: A healthcare organization implemented a structured escalation path for AI incidents, allowing employees to report issues anonymously and ensuring that concerns are addressed by the appropriate governance committee.

Impact of Clear Accountability and Governance Structures

By establishing clear roles and responsibilities, organizations can ensure that AI systems are managed effectively, risks are identified and addressed in a timely manner, and governance activities are coordinated across the enterprise.

6. Educating and Training Employees on Responsible AI Use

Overview

Employee training and education are critical components of a robust AI governance strategy. As AI technologies become more prevalent, all employees—not just data scientists—must understand the risks and ethical considerations associated with AI use. Training programs should be tailored to different roles, from technical staff to business leaders, ensuring that everyone is equipped to identify and respond to potential misuse scenarios.

Key Strategies for Educating and Training Employees

1. **Develop Role-Specific Training Programs**
   - Create customized training programs based on the roles and responsibilities of employees. For example, data scientists should receive training on ethical AI design and bias mitigation, while business leaders should focus on AI strategy and governance.
   - **Example**: A leading e-commerce firm developed separate training modules for its data scientists, legal teams, and business executives, covering topics such as AI bias, compliance, and ethical decision-making.

2. **Incorporate Real-World Case Studies into Training**
    - Use real-world case studies to illustrate the potential risks and consequences of AI misuse. Highlight examples of both successful and failed AI deployments to provide practical insights into responsible AI use.
    - **Example**: A financial services firm incorporated real-world case studies into its AI ethics training program, using past incidents of algorithmic discrimination and data breaches to emphasize the importance of governance and ethical design.
3. **Establish a Continuous Learning Program**
    - Implement a continuous learning program that keeps employees up-to-date on the latest AI governance trends, regulatory changes, and emerging risks. Use workshops, webinars, and hands-on training sessions to reinforce key concepts.
    - **Example**: A healthcare company launched a continuous learning initiative for its AI teams, offering quarterly workshops on topics such as AI fairness, data privacy, and new compliance requirements.
4. **Create a Feedback Mechanism for AI Concerns**
    - Establish a feedback mechanism that allows employees to report concerns about AI misuse or suggest improvements to governance practices. Use surveys, anonymous hotlines, or suggestion boxes to collect feedback and make continuous improvements.
    - **Example**: A multinational organization created an anonymous feedback portal where employees could raise concerns about AI projects or suggest improvements to the company's AI ethics policies.

Impact of Employee Training and Education

Training and educating employees on responsible AI use empowers them to identify and address potential risks, fosters a culture of ethical AI development, and ensures that all stakeholders are aligned with the organization's AI governance objectives.

7. Leveraging AI for AI Governance

Overview

AI itself can be used to enhance governance activities by automating compliance monitoring, detecting anomalies, and providing insights into model performance and behavior. Using AI to govern AI allows organizations to scale their governance efforts, reduce manual workloads, and respond to risks more effectively. AI-powered governance tools can detect patterns that humans might miss, making them invaluable for managing complex AI systems.

Key Strategies for Leveraging AI for AI Governance

1. **Implement AI-Based Compliance Monitoring Tools**
    - Use AI models to monitor compliance with internal policies and external regulations. These tools can analyze data usage, track model decisions, and detect potential violations in real-time.

- **Example**: A global bank implemented an AI-based compliance monitoring tool to track the use of sensitive customer data in its AI models, ensuring compliance with GDPR and other data privacy regulations.

2. **Use AI for Anomaly Detection and Risk Management**
   - Deploy AI models to detect anomalies in model behavior, such as unusual predictions or data drift, that could indicate a security issue or model misuse.
   - **Example**: A logistics company uses an AI-based anomaly detection tool to monitor its supply chain optimization models, detecting unusual patterns that could indicate manipulation or data poisoning.
3. **Automate Policy Enforcement with AI:**
   - Use AI to automatically enforce governance policies, such as data access controls, model usage restrictions, and compliance requirements. Implement AI tools that can dynamically adjust policies based on real-time conditions.
   - **Example**: A healthcare provider uses AI to enforce policies on the use of sensitive patient data in its AI models, ensuring that models are only used for approved research purposes.
4. **Use AI to Enhance Model Explainability and Transparency:**
   - Implement AI tools like **Google Explainable AI** or **Azure InterpretML** to provide insights into how models make decisions, enhancing transparency and trust in AI systems.
   - **Example**: A technology company uses AI-based explainability tools to generate detailed reports on model decisions, helping business leaders understand the rationale behind AI-driven recommendations.

Impact of Using AI for Governance

Using AI to govern AI enables organizations to scale their governance efforts, enhance the effectiveness of compliance monitoring, and respond to risks more proactively.

Conclusion

Misuse of Generative AI poses significant risks to enterprises, ranging from financial loss and reputational damage to legal liabilities. As AI technology evolves, so do the tactics of malicious actors. It is imperative for organizations to implement comprehensive governance frameworks, robust security measures, and continuous monitoring to safeguard against these threats. By understanding the misuse scenarios and adopting proactive protection strategies, enterprises can leverage Generative AI safely and responsibly.

Protecting against the misuse of Generative AI requires a holistic approach that encompasses governance, security, ethical guidelines, and employee education. By establishing comprehensive AI governance frameworks, implementing robust security measures, and leveraging AI tools for monitoring and compliance, enterprises can safeguard against the risks associated with Generative AI technologies. Continuous monitoring, training, and clear accountability structures are essential for maintaining the integrity, security, and ethical alignment of AI systems.

By adopting these strategies, organizations can ensure that their AI initiatives are not only innovative and effective but also responsible, compliant, and aligned with their core values.

Chapter 16

Generative AI Risk Categorizing

As Generative AI (GenAI) solutions continue to expand across industries, it is essential to categorize use cases by their risk profiles. This categorization helps organizations prioritize safety, ethical standards, and compliance based on the potential impact and sensitivity of the AI's application. By examining factors such as outcome impact, data sensitivity, ethical implications, and regulatory considerations, GenAI use cases can be classified into high-risk, medium-risk, and low-risk categories. This chapter explores each category in detail and introduces tools available from AWS, Google Cloud, and Azure to aid in risk assessment and categorization.

Factors for Determining Risk Categories

To accurately assign risk levels to GenAI use cases, we must evaluate them on various parameters that define potential impact and sensitivity. Key factors include:

1. Outcome Impact: How critical or irreversible are the consequences of errors in this use case?
2. Data Sensitivity: Does the use case handle sensitive or confidential data (e.g., personally identifiable information, health data)?
3. Likelihood of Ethical Concerns: Could the use case lead to biases, discrimination, or unethical behavior if not properly monitored?
4. Regulatory and Compliance Requirements: Is the use case subject to industry regulations or legal restrictions?
5. Misuse Potential: How likely is it that the AI solution could be misused or manipulated for unintended purposes?

Each use case can then be mapped into high-risk, medium-risk, or low-risk categories based on these factors, allowing for targeted risk management strategies and appropriate safeguards.

High-Risk Use Cases

Characteristics of High-Risk Use Cases

High-risk GenAI use cases have substantial impacts, often involving critical or sensitive areas where mistakes or biases can result in severe consequences. These cases are often subject to rigorous regulatory standards, involve highly sensitive data, and demand advanced measures for security, bias mitigation, and human oversight.

Key Use Cases in High-Risk Category

1. Healthcare Diagnostics and Treatment Recommendations: GenAI models used to assist with medical diagnoses, drug prescriptions, or treatment plans. Errors could lead to incorrect treatments or severe health risks.
2. Financial Decision-Making: Use cases that influence loan approvals, credit scores, or investments. Biases or inaccuracies could lead to financial harm, discrimination, or unfair lending practices.
3. Autonomous Decision-Making in Defense: AI used in autonomous weapons or critical military decision-making. Malfunctions or unintended outputs could have catastrophic consequences.
4. Legal and Judicial Applications: Models used for predicting legal case outcomes or advising sentencing. Unintended biases could lead to discrimination and impact human rights.
5. Advanced Content Generation for Political and News Media: AI generating political messages, speeches, or news articles can spread misinformation if not monitored, impacting public opinion and trust.

AWS, Google Cloud, and Azure Tools for High-Risk Categorization

- AWS: Amazon SageMaker Clarify offers tools for detecting and mitigating bias and assessing data sensitivity, while AWS Identity and Access Management (IAM) enables stringent access controls to secure sensitive data.
- Google Cloud: Explainable AI and What-If Tool in Vertex AI help ensure model transparency and detect biases, while Cloud DLP assists with data protection in sensitive applications.
- Azure: Fairlearn and InterpretML in Azure Machine Learning provide tools for bias detection and explainability, and Azure Key Vault secures sensitive information used in high-risk applications.

Medium-Risk Use Cases

Characteristics of Medium-Risk Use Cases

Medium-risk GenAI use cases have a significant but manageable impact. These applications are often less sensitive in nature or are not as tightly regulated. Errors or biases may lead to inconvenience or reduced effectiveness but are generally reversible or correctable. Ethical considerations are present but not as pronounced as in high-risk cases.

Key Use Cases in Medium-Risk Category

1. Customer Service Chatbots for General Inquiries: AI-powered bots for routine customer support queries where errors are manageable and pose little risk to user well-being.

2. Marketing and Personalization: Tailoring content, ads, or recommendations based on user profiles. Ethical considerations exist around data privacy, but consequences are typically less severe.
3. Education and Tutoring Assistants: GenAI solutions that help students learn new topics or prepare for exams. Missteps may lead to misunderstandings but are generally low-impact and correctable.
4. Content Moderation: AI models that identify and remove inappropriate content on social platforms. Mistakes may result in under- or over-moderation but do not usually have severe repercussions.
5. Employee Training and Onboarding Support: GenAI solutions used to guide new employees through company processes. Errors may slow down onboarding but rarely lead to critical issues.

AWS, Google Cloud, and Azure Tools for Medium-Risk Categorization

- AWS: Amazon Personalize for tailored recommendations, with Amazon Macie and Amazon KMS for data security to address privacy concerns.
- Google Cloud: Dialogflow for creating chatbots with Cloud IAM and Data Loss Prevention for managing and securing customer interactions.
- Azure: Azure Bot Services and Azure Personalizer enable personalization and conversational AI, with Azure Policy for compliance with data privacy standards in medium-risk environments.

Low-Risk Use Cases

Characteristics of Low-Risk Use Cases

Low-risk GenAI use cases generally involve minimal potential for harm and operate in environments where the consequences of errors are manageable and reversible. They involve non-sensitive data, limited ethical implications, and do not typically fall under strict regulatory requirements. These solutions are often focused on enhancing productivity or improving aesthetic content.

Key Use Cases in Low-Risk Category

1. Content Generation for Non-Sensitive Marketing: Creating blog posts, social media updates, or product descriptions where errors can be easily corrected and carry minimal risk.
2. Data Augmentation for Training: Generating synthetic data for training or testing AI models in environments where data sensitivity is not a concern.
3. Grammar and Writing Assistance Tools: AI-based tools for grammar checks, summarization, or rephrasing of text. Mistakes are minor and easily fixed by the user.
4. Image Enhancement and Editing: Tools for improving visual aesthetics, adjusting colors, or resizing images. Errors in outputs are non-critical.
5. Internal Knowledge Base and Document Search: GenAI-powered tools for indexing and searching non-sensitive internal documents, such as employee FAQs.

AWS, Google Cloud, and Azure Tools for Low-Risk Categorization

- AWS: Amazon Comprehend for natural language understanding in knowledge management, with Amazon S3 for secure, low-cost storage of non-sensitive content.
- Google Cloud: Cloud Vision API for image enhancement, AutoML Natural Language for internal document processing, and Cloud Storage for managing non-sensitive data.
- Azure: Azure Cognitive Search for internal document management and Azure Cognitive Services for Language for text processing, with Blob Storage for general content storage.

Parameters and Methodology for Risk Categorization

To systematically classify GenAI use cases by risk, organizations can establish a framework that scores each use case across key parameters. This quantitative approach ensures a balanced view of all contributing risk factors.

Key Parameters for Risk Scoring

1. Outcome Impact Score: Evaluate the severity of consequences in the event of an AI error. Assign higher scores to use cases where errors could lead to significant harm.
2. Data Sensitivity Score: Assess the level of data sensitivity involved in the use case. Higher scores should be assigned to cases involving PII, health, or financial data.
3. Ethical Sensitivity Score: Determine the likelihood of ethical concerns such as bias or discrimination. Use cases with greater potential for ethical risks should receive higher scores.
4. Compliance Requirement Score: Assess the degree to which the use case is subject to regulations. Use cases with stringent compliance needs, such as those in finance or healthcare, receive higher scores.
5. Misuse Potential Score: Analyze the ease with which the use case could be misused, leading to unintended harm. Assign higher scores to cases with greater misuse potential.

Based on the cumulative score, use cases can be categorized as:

- High-Risk: Total score is above a specified threshold, indicating a need for advanced risk mitigation and human oversight.
- Medium-Risk: Scores in an intermediate range, warranting some risk controls but with manageable consequences.
- Low-Risk: Lower scores, suggesting minimal risk controls are required due to low impact and sensitivity.

AWS, Google Cloud, and Azure Tools for Risk Assessment

- AWS: AWS Cloud Compliance tools support audits and regulatory assessments, while Amazon SageMaker Clarify aids in detecting bias and evaluating ethical risks.
- Google Cloud: Vertex AI Explainable AI provides insights into model predictions, supporting ethical sensitivity assessment, while Cloud DLP assists in evaluating data sensitivity.
- Azure: Azure Policy and Compliance Manager help manage compliance needs, while Fairlearn and InterpretML are used to monitor ethical aspects in the solution.

Implementing Safeguards Based on Risk Categories

Once categorized, GenAI use cases can be managed with appropriate safeguards:

1. High-Risk Use Cases:
   - Implement stringent access controls, bias mitigation tools, and regular audits.
   - Enable human oversight for critical outputs.
   - Use advanced compliance tools to ensure adherence to regulatory standards.
2. Medium-Risk Use Cases:
   - Establish periodic monitoring and feedback loops to catch potential issues early.
   - Incorporate disclaimers for users on AI limitations.
   - Use compliance policies to protect privacy and ethical standards.
3. Low-Risk Use Cases:
   - Maintain basic security and monitoring to ensure reliable performance.
   - Use flexible, low-cost resources to support scalability with minimal maintenance.
   - Monitor performance and user feedback periodically for minor improvements.

Conclusion

The systematic categorization of GenAI use cases by risk levels (high, medium, low) ensures that AI applications are responsibly managed according to their potential impact, sensitivity, and ethical implications. By implementing tailored safeguards based on risk profiles, organizations can deploy Generative AI solutions that are secure, compliant, and aligned with ethical standards. Leveraging tools from AWS, Google Cloud, and Azure further enhances the organization's ability to assess, categorize, and manage risks effectively, enabling responsible AI deployment at scale. Below is the summarization table

Risk Level	Characteristics	Examples	Mitigation
High Risk	High-stakes decisions, sensitive data, severe consequences	Healthcare diagnosis, legal decisions, military applications, hiring, critical customer support	Human oversight, compliance with regulations, bias detection, explainability
Medium Risk	Moderate impact, ethical considerations, manageable consequences	Marketing personalization, chatbots, automated content moderation, tutoring assistants	Monitoring, clear disclaimers, anonymized data, regular audits
Low Risk	Minimal impact, low sensitivity, limited harm potential	Content creation, data augmentation, personal productivity tools, image enhancement	Basic monitoring, minimal oversight, general disclaimers

Chapter 17

The Role of the Chief Generative AI Officer

With the rapid adoption of Generative AI across industries, there is an emerging need for a dedicated leadership role that oversees AI strategy, implementation, and governance. The **Chief Generative AI Officer (CGO)** is a strategic position designed to guide organizations in leveraging Generative AI technologies to drive innovation, enhance efficiency, and maintain a competitive edge. This chapter explores the responsibilities, skill sets, and strategic impact of a Chief Generative AI Officer, highlighting how this role aligns AI initiatives with broader business objectives while ensuring the ethical and responsible use of AI technologies.

1. Understanding the Role of a Chief Generative AI Officer (CGO)

The Chief Generative AI Officer is a senior executive responsible for defining and executing the organization's Generative AI strategy. Unlike traditional Chief Data Officers (CDOs) or Chief Technology Officers (CTOs), the CGO focuses specifically on harnessing the capabilities of Generative AI models—such as Large Language Models (LLMs), image generators, and text-to-speech systems—to create transformative business solutions.

Why This Role is Important:

The advent of advanced Generative AI models has shifted the focus from merely using AI for predictive analytics to generating new content, automating creative processes, and augmenting human capabilities. This shift requires a specialized role that not only understands the technical intricacies of these models but also their strategic business impact. The CGO is responsible for bridging the gap between AI research and business value, ensuring that AI initiatives are not just innovative but also aligned with the organization's long-term goals.

2. Key Responsibilities of a Chief Generative AI Officer

The CGO's role encompasses a broad range of responsibilities, from strategic planning and governance to operational oversight and risk management. Below are the primary responsibilities of a CGO, along with practical examples of how they contribute to the organization.

2.1. Defining and Driving the Generative AI Strategy

Description:

The CGO is responsible for defining the Generative AI vision and strategy for the organization. This involves identifying key areas where Generative AI can create value, setting long-term goals, and developing a roadmap for AI integration across various business units.

Why It's Important:

A clear AI strategy ensures that Generative AI initiatives are aligned with the organization's overall business strategy, focusing on high-impact areas and avoiding fragmented or siloed efforts.

Example:

In a retail organization, the CGO might develop a strategy to use Generative AI for personalized marketing, virtual try-on experiences, and automated content generation for e-commerce websites. This strategic direction helps prioritize AI projects that directly contribute to revenue growth and customer satisfaction.

Key Activities:

- Conduct AI maturity assessments to identify current capabilities and gaps.
- Develop a comprehensive Generative AI strategy document that outlines priorities, resources, and timelines.
- Work with the Chief Data Officer (CDO) and Chief Technology Officer (CTO) to align AI strategy with broader data and IT initiatives.

2.2. Building and Leading the Generative AI Team

Description:

The CGO is responsible for building a cross-functional Generative AI team, including data scientists, AI researchers, engineers, and domain experts. This team collaborates to develop, deploy, and optimize Generative AI models tailored to the organization's needs.

Why It's Important:

A strong AI team with diverse expertise is crucial for developing high-quality models, overcoming technical challenges, and ensuring that the AI systems are both performant and aligned with business goals.

Example:

A CGO at a financial services company might establish a Generative AI team focused on automating complex document analysis, creating personalized financial reports, and assisting in fraud detection using text and speech generation technologies.

Key Activities:

- Define roles and responsibilities within the Generative AI team.
- Recruit top talent in AI research, data engineering, and domain-specific fields.
- Foster a culture of innovation, continuous learning, and ethical AI development.

2.3. Governance and Ethical Oversight

Description:

The CGO ensures that all Generative AI initiatives comply with ethical standards, industry regulations, and internal governance policies. This involves establishing frameworks for responsible AI usage, monitoring AI deployments for compliance, and setting up mechanisms to handle ethical dilemmas.

Why It's Important:

Generative AI models, by their nature, can produce outputs that are unpredictable or even harmful if not properly governed. Ensuring ethical AI usage is critical to maintaining trust, avoiding regulatory penalties, and protecting the organization's reputation.

Example:

In a healthcare organization, the CGO might implement a governance framework that prevents AI models from generating biased or misleading medical advice. This could include regular audits, bias detection tools, and a dedicated ethics committee.

Key Activities:

- Develop and enforce AI governance frameworks that include data privacy, security, and bias mitigation.
- Implement tools for monitoring and auditing AI models to detect and rectify unethical behavior.
- Create an internal ethics board to review and approve high-impact Generative AI projects.

2.4. Managing AI Risks and Compliance

Description:

The CGO is responsible for identifying and managing the risks associated with Generative AI models, such as model bias, data security, compliance with regulations like GDPR, and risks of misinformation or misuse of AI outputs.

Why It's Important:

Managing AI risks proactively prevents potential legal liabilities, financial losses, and reputational damage. A well-defined risk management strategy ensures that the organization's AI deployments are safe, reliable, and compliant.

Example:

A CGO in a global insurance company might deploy tools like AWS SageMaker Clarify or Google Cloud's What-If Tool to detect biases in underwriting models and ensure compliance with local data privacy regulations.

Key Activities:

- Implement AI risk management frameworks to identify and mitigate risks early.
- Set up compliance monitoring systems that align with industry regulations.
- Conduct regular risk assessments for AI models, especially those used in critical or regulated sectors.

2.5. Driving Innovation and Research

Description:

The CGO fosters a culture of innovation by promoting research and development (R&D) in the field of Generative AI. This includes exploring emerging technologies, experimenting with new AI models, and driving the adoption of cutting-edge research to maintain a competitive edge.

Why It's Important:

Generative AI is a rapidly evolving field, and staying ahead of the curve is essential for maintaining a competitive advantage. A focus on innovation enables the organization to capitalize on the latest advancements and explore new business opportunities.

Example:

A CGO in a media company might lead R&D efforts to explore how Generative AI can be used for automated video production, content generation, and deepfake detection, creating new revenue streams and enhancing operational efficiency.

Key Activities:

- Set up innovation labs or centers of excellence dedicated to Generative AI.
- Collaborate with academic institutions and research organizations to stay updated on the latest AI advancements.
- Allocate resources for exploratory projects that push the boundaries of what Generative AI can achieve.

3. Skills and Qualifications for a Chief Generative AI Officer

To successfully fulfill the responsibilities outlined above, a CGO must possess a unique combination of technical expertise, strategic vision, and leadership skills. Below are the key skills and qualifications required for this role:

1. **Technical Proficiency:**
    - Deep understanding of Generative AI models, including transformers, GANs, and variational autoencoders (VAEs).
    - Experience with AI platforms such as **AWS SageMaker**, **Azure Machine Learning**, and **Google Vertex AI**.
2. **Strategic Vision:**
    - Ability to align AI initiatives with business goals and identify high-impact areas for Generative AI adoption.
3. **Leadership and Team Management:**
    - Experience building and leading cross-functional AI teams, fostering a collaborative and innovative culture.
4. **Ethical and Governance Expertise:**
    - Knowledge of AI ethics, data privacy regulations, and compliance frameworks.

4. Conclusion

The Chief Generative AI Officer is a critical role for organizations looking to harness the full potential of Generative AI. By defining strategy, building high-performing teams, ensuring governance and compliance, and driving innovation, the CGO ensures that AI initiatives deliver tangible business value while upholding ethical standards. As Generative AI continues to evolve, the CGO will play a pivotal role in shaping how organizations use AI to transform their operations, create new products, and engage with customers in innovative ways.

Chapter 18

Comprehensive List of Generative AI Use Cases Across All Industries

Generative AI (GenAI) has emerged as a transformative technology across a broad spectrum of industries, offering innovative solutions that enhance productivity, improve decision-making, and create new opportunities for automation and creativity. From healthcare to finance, manufacturing to entertainment, GenAI models are being used to generate human-like text, create realistic images, simulate complex systems, and automate repetitive tasks. This chapter presents a comprehensive list of Generative AI use cases organized by industry, showcasing the diverse applications of this powerful technology.

Healthcare and Life Sciences

1. **Medical Imaging Analysis**: Use AI models to generate detailed, annotated medical images to assist in the diagnosis of conditions such as tumors or fractures.
2. **Drug Discovery and Development**: Simulate molecular structures and predict chemical properties to accelerate drug discovery.
3. **Clinical Trial Design and Simulation**: Use AI to design optimized clinical trials, simulate patient responses, and predict trial outcomes.
4. **Electronic Health Record (EHR) Summarization**: Generate concise summaries of complex patient histories to aid in clinical decision-making.
5. **Virtual Health Assistants**: Create AI-powered chatbots to assist patients with symptom checking, appointment scheduling, and health information retrieval.
6. **Medical Report Generation**: Automatically generate radiology or pathology reports from raw data.
7. **Synthetic Data Generation for Research**: Create synthetic patient datasets to support research and model training without compromising patient privacy.

8. **Medical Literature Review**: Use AI to generate summaries and insights from large volumes of research articles and clinical studies.
9. **Personalized Health Recommendations**: Generate personalized treatment plans and lifestyle recommendations based on patient data.
10. **Genomic Data Analysis**: Use AI to simulate gene expressions and generate insights for precision medicine.

Financial Services

1. **Algorithmic Trading**: Generate trading strategies and simulate market scenarios for automated trading systems.
2. **Risk Assessment and Fraud Detection**: Use AI models to generate patterns for detecting fraudulent transactions and assessing credit risk.
3. **Financial Report Generation**: Automate the creation of quarterly reports, earnings summaries, and investment recommendations.
4. **Personalized Financial Advisory**: Generate personalized investment strategies and financial plans for clients based on their profiles.
5. **Market Sentiment Analysis**: Analyze and generate insights from financial news and social media to predict market movements.
6. **Insurance Claim Processing**: Automate the generation and processing of insurance claims, including document verification and damage estimation.
7. **Synthetic Data for Model Training**: Create synthetic financial datasets to test and validate trading algorithms.
8. **Predictive Maintenance for ATMs**: Use AI to predict when ATMs will require maintenance based on historical usage patterns.
9. **Customer Support Automation**: Implement AI chatbots to generate responses for customer inquiries, such as loan applications or account management.
10. **Portfolio Optimization**: Use AI to generate and optimize investment portfolios based on risk tolerance and market conditions.

Manufacturing and Supply Chain

1. **Predictive Maintenance**: Generate maintenance schedules and simulate equipment failures to prevent unplanned downtime.
2. **Supply Chain Optimization**: Use AI to simulate supply chain disruptions and optimize inventory management and logistics.
3. **Product Design and Prototyping**: Generate 3D models and simulate product performance for rapid prototyping.
4. **Quality Control Automation**: Use AI to generate defect detection patterns for automated quality inspections.
5. **Demand Forecasting**: Generate demand forecasts based on historical data, market trends, and seasonal patterns.

6. **Robotics and Process Automation**: Implement AI to generate and optimize robotic workflows for manufacturing processes.
7. **Inventory Management**: Use AI to generate reorder recommendations and optimize stock levels.
8. **Digital Twin Simulation**: Create digital twins of manufacturing systems to simulate and optimize production lines.
9. **Supply Chain Risk Analysis**: Generate risk scenarios and predict the impact of disruptions on supply chain operations.
10. **Energy Consumption Optimization**: Simulate energy usage and generate recommendations for reducing operational costs.

Retail and E-Commerce

1. **Personalized Product Recommendations**: Use AI to generate personalized product suggestions based on customer behavior and preferences.
2. **Dynamic Pricing Optimization**: Simulate and generate optimal pricing strategies for maximizing revenue.
3. **Customer Sentiment Analysis**: Analyze and generate insights from customer reviews and feedback to improve products and services.
4. **Chatbots for Customer Support**: Implement AI chatbots to generate responses for customer inquiries and support.
5. **Virtual Shopping Assistants**: Create AI-powered virtual assistants that guide customers through product selections.
6. **Inventory and Supply Chain Management**: Generate forecasts and optimize inventory management to minimize stockouts and overstock.
7. **Product Content Generation**: Automate the generation of product descriptions, ad copy, and promotional content.
8. **Visual Search and Product Matching**: Implement AI to generate visual search capabilities and find similar products based on images.
9. **Customer Experience Simulation**: Use AI to simulate different customer experiences and optimize store layouts.
10. **Loyalty Program Optimization**: Generate personalized loyalty offers and engagement strategies based on customer data.

Media and Entertainment

1. **Content Creation and Story Generation**: Use AI to generate storylines, scripts, and content ideas for movies, games, and books.
2. **Realistic Character Animation**: Create realistic character animations for movies and video games.
3. **Image and Video Synthesis**: Use AI to generate high-quality images, video clips, and special effects.
4. **Music Composition**: Generate original music tracks based on genre, mood, and artist preferences.
5. **Voice Synthesis and Dubbing**: Implement AI to generate synthetic voices for dubbing and voiceovers.

6. **Automated Video Editing**: Generate edits and transitions automatically based on scene recognition and content analysis.
7. **Recommendation Systems**: Use AI to generate personalized content recommendations for streaming services.
8. **Real-Time Transcription and Translation**: Implement AI for real-time subtitle generation and language translation.
9. **News Article Summarization**: Automatically generate summaries of news articles and video content.
10. **Sentiment Analysis for Media Content**: Use AI to analyze and generate sentiment insights for social media posts and comments.

Legal and Compliance

1. **Contract Generation and Review**: Use AI to automatically generate and review contracts, identifying risks and suggesting improvements.
2. **Legal Research and Case Analysis**: Generate summaries and insights from legal texts, case law, and regulations.
3. **Compliance Monitoring**: Automate the generation of compliance reports and monitor regulatory changes.
4. **Intellectual Property Management**: Use AI to generate and review patent applications, copyright filings, and trademark documents.
5. **Document Summarization**: Automatically generate summaries of lengthy legal documents and agreements.
6. **Risk Assessment and Due Diligence**: Implement AI to generate risk assessments and perform due diligence for mergers and acquisitions.
7. **Legal Chatbots**: Use AI to generate responses for common legal inquiries, such as employment law or contract disputes.
8. **Case Outcome Prediction**: Generate predictions for case outcomes based on historical data and legal precedents.
9. **Data Privacy Management**: Use AI to generate and enforce data privacy policies based on regulatory requirements.
10. **Regulatory Impact Analysis**: Generate impact analysis reports for proposed regulatory changes on existing business practices.

Education and Learning

1. **Personalized Learning Pathways**: Generate personalized learning plans and course recommendations for students.
2. **Automated Essay Scoring**: Use AI to generate essay evaluations and feedback based on predefined rubrics.
3. **Content Creation for Courses**: Automate the generation of educational content, including quizzes, tutorials, and study guides.
4. **Virtual Tutors**: Create AI-powered virtual tutors to assist students with course materials and homework.

5. **Language Learning Assistants**: Use AI to generate language learning exercises and real-time pronunciation feedback.
6. **Educational Chatbots**: Implement chatbots to answer common questions and guide students through learning modules.
7. **Interactive Simulations and Scenarios**: Use AI to generate interactive learning scenarios for subjects like history, science, or mathematics.
8. **Student Performance Analytics**: Generate predictive analytics to identify at-risk students and recommend intervention strategies.
9. **Gamified Learning Experiences**: Use AI to generate game-based learning content to engage students.
10. **Automated Certificate Generation**: Automatically generate course completion certificates and badges.

Automotive and Transportation

1. **Autonomous Vehicle Simulation**: Use AI to generate realistic driving scenarios for training and testing autonomous vehicles.
2. **Predictive Maintenance**: Generate maintenance schedules and simulate vehicle wear and tear to prevent mechanical failures.
3. **Route Optimization**: Use AI to generate the most efficient routes for logistics and transportation, considering traffic patterns and weather conditions.
4. **Driver Behavior Analysis**: Implement AI to analyze and generate insights on driver behavior for safety improvements.
5. **Fleet Management**: Automate the generation of fleet management strategies, including vehicle allocation and fuel optimization.
6. **Vehicle Design and Prototyping**: Use AI to generate vehicle designs and simulate aerodynamics for better fuel efficiency.
7. **Traffic Management**: Implement AI to generate optimized traffic control strategies and predict congestion patterns.
8. **Passenger Experience Personalization**: Use AI to generate personalized in-vehicle experiences, such as customized entertainment or navigation suggestions.
9. **Supply Chain Logistics**: Use AI to simulate and optimize supply chain logistics for automotive components and materials.
10. **Accident Analysis and Reconstruction**: Use AI to simulate and reconstruct traffic accidents for forensic analysis and safety research.

Energy and Utilities

1. **Energy Demand Forecasting**: Use AI to generate predictions for energy consumption based on historical data and weather patterns.
2. **Smart Grid Optimization**: Generate optimal grid management strategies to balance supply and demand in real time.
3. **Predictive Maintenance for Power Plants**: Use AI to monitor equipment and generate early warning alerts for potential failures.

4. **Energy Efficiency Optimization**: Generate recommendations for improving energy efficiency in industrial and residential settings.
5. **Renewable Energy Integration**: Simulate and optimize the integration of renewable energy sources, such as solar and wind, into existing grids.
6. **Oil and Gas Exploration**: Use AI to generate geological models and predict the likelihood of resource deposits.
7. **Utility Billing Automation**: Automate the generation of detailed utility bills and consumption reports for customers.
8. **Energy Trading Optimization**: Use AI to generate strategies for trading energy on open markets.
9. **Carbon Footprint Analysis**: Generate real-time carbon footprint reports and suggest ways to reduce emissions.
10. **Disaster Management and Recovery**: Simulate natural disasters and generate response plans to ensure energy continuity.

Telecommunications

1. **Network Optimization**: Use AI to generate optimized network configurations for minimizing latency and maximizing bandwidth.
2. **Customer Support Automation**: Implement AI-powered chatbots to generate responses for common technical issues and service inquiries.
3. **Churn Prediction and Retention Strategies**: Generate predictive models for customer churn and suggest retention strategies.
4. **Fraud Detection**: Use AI to identify and generate patterns for detecting telecom fraud, such as SIM swapping or phishing.
5. **Service Provisioning Automation**: Automate the generation of service provisioning plans for new customer installations.
6. **Dynamic Bandwidth Allocation**: Use AI to generate strategies for dynamically allocating bandwidth based on user demand.
7. **Network Security Monitoring**: Generate alerts and recommendations for responding to network security threats.
8. **5G Network Planning**: Use AI to simulate and generate optimal deployment strategies for 5G network infrastructure.
9. **Call Quality Analysis**: Analyze and generate insights on call quality to optimize service delivery.
10. **Content Delivery Network (CDN) Optimization**: Use AI to simulate and optimize CDN configurations for better content delivery.

Real Estate and Property Management

1. **Property Price Prediction**: Use AI to generate price predictions based on market trends, property features, and neighborhood data.
2. **Virtual Property Tours**: Create AI-generated virtual tours and 3D models of properties for remote viewings.

3. **Tenant Screening and Risk Assessment**: Use AI to generate tenant profiles and assess risks based on credit history and rental behavior.
4. **Property Listing Automation**: Automate the generation of property listings, including descriptions, images, and pricing details.
5. **Smart Building Management**: Use AI to optimize heating, ventilation, and air conditioning (HVAC) systems based on occupancy patterns.
6. **Investment Analysis and Forecasting**: Generate forecasts for property investments and simulate different market scenarios.
7. **Lease Management Automation**: Automate the generation and review of lease agreements and renewals.
8. **Real Estate Portfolio Optimization**: Use AI to generate strategies for optimizing real estate portfolios based on risk and return.
9. **Customer Engagement Bots**: Implement AI-powered bots to handle inquiries from prospective buyers and renters.
10. **Urban Planning and Development**: Use AI to simulate urban development scenarios and generate recommendations for land use.

Agriculture and Farming

1. **Crop Yield Prediction**: Use AI to generate crop yield predictions based on weather patterns, soil conditions, and historical data.
2. **Precision Agriculture**: Generate recommendations for planting, irrigation, and fertilization to optimize crop production.
3. **Pest and Disease Detection**: Use AI to generate alerts for early detection of pests and diseases using satellite or drone imagery.
4. **Livestock Health Monitoring**: Implement AI to generate health reports and early warning alerts for livestock diseases.
5. **Smart Irrigation Systems**: Use AI to generate optimized irrigation schedules based on weather forecasts and soil moisture levels.
6. **Supply Chain Management**: Generate supply chain optimization strategies for transporting and distributing agricultural products.
7. **Climate Impact Simulation**: Simulate the impact of climate change on crop yields and generate adaptation strategies.
8. **Farm Automation**: Use AI to automate machinery for planting, harvesting, and soil preparation.
9. **Food Waste Reduction**: Generate recommendations for reducing food waste along the supply chain.
10. **Sustainable Farming Practices**: Use AI to generate insights on sustainable farming techniques, such as crop rotation and organic farming.

Travel and Hospitality

1. **Personalized Travel Itinerary Generation**: Use AI to generate personalized travel itineraries based on user preferences and interests.
2. **Dynamic Pricing for Hotel Rooms**: Implement AI to generate optimized pricing strategies based on demand, seasonality, and competitor pricing.
3. **Customer Sentiment Analysis**: Use AI to analyze and generate insights from customer reviews and feedback.
4. **Virtual Travel Assistants**: Create AI-powered virtual assistants to guide customers through booking processes and provide travel information.
5. **Predictive Maintenance for Hotel Equipment**: Use AI to generate maintenance schedules for HVAC systems and other equipment in hotels.
6. **Guest Experience Personalization**: Generate personalized recommendations for hotel services, such as spa treatments and dining options.
7. **Travel Demand Forecasting**: Use AI to predict travel demand and optimize inventory management for airlines and hotels.
8. **Automated Travel Document Processing**: Use AI to automate the generation and verification of travel documents, such as visas and passports.
9. **Chatbots for Customer Service**: Implement AI chatbots to generate responses for booking inquiries and provide travel assistance.
10. **Loyalty Program Optimization**: Use AI to generate personalized offers and engagement strategies for frequent travelers.

Government and Public Sector

1. **Policy Impact Analysis**: Use AI to simulate and generate impact assessments for proposed policies and regulations.
2. **Fraud Detection and Prevention**: Implement AI models to generate fraud detection patterns in areas such as tax filings, social benefits, and public procurements.
3. **Smart City Planning**: Use AI to generate optimized urban planning strategies for transportation, energy usage, and public services.
4. **Public Health Monitoring**: Generate insights on public health trends using AI to analyze disease spread and recommend intervention strategies.
5. **Automated Citizen Services**: Use AI-powered chatbots to generate responses for common citizen inquiries and automate service requests.
6. **Resource Allocation and Budgeting**: Use AI to generate optimized budget allocations and resource management strategies.
7. **Disaster Response and Management**: Implement AI to simulate disaster scenarios and generate emergency response plans.
8. **Crime Prediction and Analysis**: Use AI to generate crime pattern analyses and predict high-risk areas to optimize police patrolling.
9. **Document Classification and Automation**: Automate the classification and processing of large volumes of government documents and records.

10. **Data Privacy and Security Management**: Use AI to generate risk assessments for data privacy and enforce compliance with regulations.

Human Resources and Workforce Management

1. **Talent Acquisition and Screening**: Use AI to generate candidate evaluations and match profiles to job descriptions based on resumes and cover letters.
2. **Employee Engagement and Sentiment Analysis**: Implement AI to generate insights from employee feedback and monitor workplace sentiment.
3. **Performance Management**: Use AI to generate personalized performance review templates and development plans based on employee behavior.
4. **Workforce Scheduling Optimization**: Generate optimized work schedules and shift assignments based on employee availability and demand forecasts.
5. **Career Path Recommendations**: Use AI to generate career development plans and suggest training opportunities for employees.
6. **Employee Churn Prediction**: Implement AI to generate predictions on employee turnover and suggest retention strategies.
7. **Compensation and Benefits Analysis**: Use AI to analyze and generate compensation benchmarks based on industry standards and performance metrics.
8. **Diversity and Inclusion Analysis**: Generate insights on workforce diversity and recommend strategies for creating a more inclusive environment.
9. **Onboarding Automation**: Automate the generation of onboarding materials and orientation schedules for new hires.
10. **Workplace Health and Safety Monitoring**: Use AI to generate alerts and recommendations for improving workplace safety.

Education and Learning

1. **Personalized Learning Pathways**: Generate customized learning plans for students based on their strengths, weaknesses, and learning preferences.
2. **AI-Generated Tutoring Systems**: Implement AI to simulate interactive tutoring sessions and generate real-time feedback for students.
3. **Virtual Classroom Assistants**: Create AI-powered assistants to support teachers by automating administrative tasks and generating content.
4. **Student Performance Prediction**: Use AI to generate predictions on student performance and suggest intervention strategies for at-risk students.
5. **Curriculum Design Optimization**: Automate the generation of curriculum designs and course structures based on educational standards.
6. **Automated Grading and Assessment**: Use AI to evaluate essays and generate assessments based on predefined rubrics.
7. **Language Learning Assistants**: Create AI-powered applications to simulate language conversations and provide real-time pronunciation feedback.

8. **Interactive Learning Content Creation**: Use AI to generate interactive and gamified learning content for subjects like math and science.
9. **Educational Resource Recommendations**: Generate personalized recommendations for textbooks, research articles, and learning materials.
10. **Career Counseling and Guidance**: Use AI to generate career recommendations and guidance based on student interests and academic performance.

Pharmaceutical and Biotechnology

1. **Drug Discovery and Molecular Simulation**: Use AI to simulate molecular interactions and generate potential drug candidates.
2. **Clinical Trial Optimization**: Generate optimized clinical trial designs and simulate patient responses for various trial scenarios.
3. **Biomarker Discovery**: Use AI to analyze and generate insights from genomic data to identify potential biomarkers for diseases.
4. **Gene Editing and CRISPR Simulation**: Implement AI to simulate CRISPR gene editing and predict outcomes.
5. **Synthetic Biology and Protein Design**: Use AI to generate novel protein structures and simulate their stability and function.
6. **Patient Stratification for Personalized Medicine**: Generate patient segments based on genomic and clinical data to tailor treatments.
7. **Drug Repurposing**: Use AI to generate insights for repurposing existing drugs for new therapeutic applications.
8. **Adverse Event Prediction**: Implement AI to predict and generate alerts for potential adverse drug reactions.
9. **Automated Regulatory Compliance**: Use AI to generate and review regulatory documentation for new drug applications.
10. **Virtual Drug Compounding**: Use AI to simulate drug compounding and optimize formulations based on patient-specific data.

Marketing and Advertising

1. **Personalized Content Generation**: Use AI to generate customized marketing messages and ad copy based on customer segments.
2. **Audience Targeting Optimization**: Generate optimal audience segments and target strategies based on demographic and behavioral data.
3. **Social Media Content Creation**: Implement AI to generate social media posts, hashtags, and engagement strategies.
4. **Sentiment Analysis for Brand Monitoring**: Use AI to analyze and generate insights from social media comments and brand mentions.
5. **Customer Journey Simulation**: Generate simulations of customer journeys to identify bottlenecks and optimize user experiences.

6. **Email Campaign Personalization**: Use AI to generate personalized email content and automate campaign scheduling.
7. **Ad Performance Prediction**: Implement AI to generate predictions on the effectiveness of ad creatives and budget allocation.
8. **Dynamic Ad Creation**: Use AI to generate dynamic ads that adapt in real-time based on user interactions and preferences.
9. **Product Name and Slogan Generation**: Generate creative product names, slogans, and taglines using AI models.
10. **Brand Voice and Style Consistency**: Use AI to analyze and generate content that aligns with the brand's voice and style guidelines.

Consumer Electronics and IoT

1. **Smart Home Automation**: Use AI to generate automation rules for smart home devices based on user behavior patterns.
2. **Voice Assistant Personalization**: Implement AI to generate personalized interactions for voice assistants based on user preferences.
3. **Predictive Maintenance for Smart Devices**: Use AI to monitor device health and generate maintenance alerts before failures occur.
4. **Energy Usage Optimization**: Generate recommendations for optimizing energy usage in smart homes and IoT-enabled buildings.
5. **Smart Device Integration and Compatibility**: Use AI to simulate and generate compatibility tests for various smart devices.
6. **Real-Time Health Monitoring**: Implement AI to generate real-time health insights from wearable devices.
7. **Smart Agriculture Monitoring**: Use AI to generate insights from IoT sensors in agriculture, such as soil moisture and crop health.
8. **Home Security Optimization**: Generate alerts and recommendations for improving home security based on sensor data.
9. **Automated Device Configuration**: Use AI to automate the setup and configuration of new smart devices.
10. **IoT Network Security Monitoring**: Implement AI to generate real-time security alerts for IoT networks.

Aerospace and Defense

1. **Mission Planning and Simulation**: Use AI to generate mission scenarios and simulate outcomes for defense operations.
2. **Flight Path Optimization**: Implement AI to generate optimized flight paths for reducing fuel consumption and improving safety.
3. **Predictive Maintenance for Aircraft**: Use AI to generate maintenance schedules and predict component failures in aircraft.

4. **Satellite Image Analysis**: Generate insights from satellite images for applications in agriculture, urban planning, and environmental monitoring.
5. **Threat Detection and Analysis**: Use AI to simulate and generate threat detection models for national security.
6. **Space Mission Simulation**: Use AI to simulate space missions and generate predictions for different scenarios.
7. **Autonomous Drone Navigation**: Implement AI to generate autonomous navigation strategies for drones in various environments.
8. **Radar and Sensor Data Analysis**: Use AI to analyze radar and sensor data to generate situational awareness insights.
9. **Combat Strategy Simulation**: Generate combat strategy simulations to train personnel and optimize tactical decisions.
10. **Automated Surveillance and Reconnaissance**: Use AI to automate surveillance data analysis and generate real-time reconnaissance insights.

This chapter offers a broad overview of how Generative AI is being applied across different industries, showcasing the versatility and transformative potential of this technology. By leveraging AI models in these diverse scenarios, organizations can optimize operations, create innovative products, and deliver more personalized services, paving the way for a smarter and more efficient future.

Acknowledgments to the Reader

Dear Reader,

Thank you for investing your time in exploring *The Gen AI Blueprint*. Writing this book has been a journey of discovery and reflection, and it is truly fulfilling to know that you chose to be a part of it. Your time and attention are precious, and I am deeply grateful that you entrusted this book to guide you through the fascinating world of Generative AI.

By the end of this book, I hope you have gained:

- **A Deep Understanding of Generative AI**: From its foundational principles to its advanced applications, you now possess the knowledge to navigate this transformative technology with confidence.
- **Practical Insights and Tools**: You've explored actionable strategies, best practices, and frameworks for implementing Generative AI solutions in real-world scenarios.
- **Awareness of Challenges and Opportunities**: Equipped with an understanding of risks, ethical considerations, and governance practices, you're ready to innovate responsibly.
- **Inspiration to Innovate**: Whether you're building a Generative AI solution, leading AI initiatives, or simply exploring its potential, I hope this book has sparked ideas and empowered you to take action.

Your journey through this book reflects your curiosity and commitment to embracing the future of AI. I hope it serves as a lasting resource and a source of inspiration as you explore the endless possibilities of Generative AI.

Thank you once again for allowing *The Gen AI Blueprint* to be a part of your journey. I encourage you to share your thoughts, feedback, and experiences—it would mean the world to me to hear how this book has contributed to your understanding and aspirations.

With gratitude,
Shyam Narayan
Author of *The Gen AI Blueprint*

Made in United States
North Haven, CT
05 November 2026

21766921R00117